Advances in Spatial and Network Economics

Titles in the Series

Martin J. Beckmann and *Tönu Puu*
Spatial Structures

Aisling J. Reynolds-Feighan
The Effects of Deregulation on U.S. Air Networks

Å. E. Andersson · D. F. Batten
K. Kobayashi · K. Yoshikawa (Eds.)

The Cosmo-Creative Society

Logistical Networks in a Dynamic Economy

With Contributions by
Åke E. Andersson, C. Anderstig, D. F. Batten, H. Haken
E. Hideshima, B. Hårsman, B. Johansson, M. Kashiwadani,
K. Kobayashi, S. Mun, N. Okada, M. Okumura, D. G. Saari,
K. Sasaki, G. J. W. Smith, S. Sunao, K. Yoshikawa, W. B. Zhang

Springer-Verlag
Berlin Heidelberg New York
London Paris Tokyo
Hong Kong Barcelona
Budapest

Prof. Dr. Åke E. Andersson
Institute for Futures Studies, Hagagatan 23B, 3tr.,
S-113 85 Stockholm, Sweden

Prof. Dr. David F. Batten
Royal Institute of Technology, Department of
Regional Planning, S-100 44 Stockholm, Sweden

Prof. Dr. Kiyoshi Kobayashi
Tottori University, Department of Social Systems
Engineering, Tottori 680, Japan

Prof. Dr. Kazuhiro Yoshikawa
Kyoto University, School of Civil Engineering,
Kyoto 606, Japan

With 67 Figures

ISBN -13:978-3-642-78462-0 e-ISBN-13:978-3-642-78460-6
DOI.10:1007/978-3-642-78460-6

Softcover reprint of the hardcover 1st edition 1993

42/7130-543210 - Printed on acid-free paper

FOREWORD

Today, telecommunication systems are expanding and evolving at a remarkable rate, with the aid of fiber optics, satellites and computerized switchboard systems. Airline systems are providing faster and more efficient networks for world-wide human transportation. Computers are now generally accessible to virtually all industries and many households. But perhaps the most important factor is that education systems are expanding the knowledge base for city populations, thus resulting in increased efficiency in the use of computers, telecommunications and rapid transportation systems.

The revolutionary age of logistical networks is upon us. Logistical networks are those systems which facilitate the movement of knowledge, commodities, money, and people in association with the production or consumption of goods and services. Logistical networks form a set of important infrastructure which serve as hard and soft means to sustain all kinds of movement, transactions and diffusion within and between global networks of cities. Major structural changes in the regional and urban economy, culture and institutions are triggered by slow but steady changes in global logistical systems.

The propensity towards the synchronized global integration of a nation's economy and society will bring about a major world-wide redistribution of production and of knowledge and information stocks, with the relative decline of the Pan-Atlantic sphere in favor of the Pan-Pacific sphere. In this new era, metropolitan areas in Japan - namely, the Tokyo Bay Area and the Osaka Bay Area, etc. - are expected to play a key role as new symbiotic centers for different cultures of the Orient and the Occident. They are simply endowed with comparative advantages in terms of access to the Asian and Pan-Pacific countries, and their multitude of traditional cultures and relationships.

The growth of creativity through cultural and artistic activities, scientific research, industrial development, administrative functions, and international communications, constitutes one of the most indispensable preconditions for a qualitative expansion of the metropolitan areas. The term 'cosmo-creative society', which is one of the central keywords in this book, refers to a creative world society playing a pioneering role in the logistical revolution occurring across global networks. It is a keyword which symbolizes the emergence of a new type of city. Scientific and practical resources should therefore be channeled towards creating symbiotic preconditions for the qualitative development of the metropolitan areas.

This volume contains a collection of papers presented at the International Forum on Logistical Dynamics and Its Regional Consequences in Osaka and the International Colloquium on Creativity and Logistical Dynamics. The Forum was held in Osaka, Japan from July 17 to 18 and the Colloquium took place in Kyoto from July 20-21 1990. The papers addressed the above-mentioned themes from the viewpoint of various interdisciplinary perspectives. The aim of the Forum was to assemble leading international scientists, administrative officials and Japanese industrial leaders in order to propose a set of scientific perspectives to facilitate an understanding of the fundamental dynamics underlying logistical revolutions and to apply this framework as a basis for determining key policy implications for the revitalization of the Osaka Bay Area. The Colloquium was organized to provide an arena for an intensive discussion of the scientific issues taken up in the Forum.

The conferences were initiated by an international research group including, among others, myself and Professor Åke E. Andersson, Director of the Institute for Futures Studies, Sweden, both of whom played a key role in promoting the conferences. I acknowledge my advisors, Professor Eiji Kometani and Professor Minoru Beika, whose unfailing encouragement and advice enabled me to organize both conferences successfully. Practical arrangements were undertaken by an Organizing Committee consisting of staff-members from the Prefectural governments of Osaka, Hyogo and Wakayama, the Municipal governments of Osaka and Kobe, The Osaka Science & Technology Center, The Osaka Industrial Association, The Osaka Chamber of Commerce & Industry, The Kansai Airport Research Institute, The Kansai Economic Federation, The Center for Industrial Innovation of Kansai, The Kansai Institute of Information Society and The Kobe Chamber of Commerce & Industry. That both conferences were successful was a tribute to the efficiency of the organizing committees and secretariats. Further I would like to express my appreciation to my friends, Professor David F. Batten, Professor Börje Johansson, Professor Kiyoshi Kobayashi, Professor Norio Okada, Dr. Seishin Sunao, and Dr. Bo Wijkmark. Without their collaboration, the conferences would not successfully end. I would also like to acknowledge the assistance provided by the Swedish Council for Building Research during the production of this book.

Last but not least, I would like to thank Hirokazu Tatano and Jennifer Wundersitz, who made invaluable contributions to the editorial process.

Kazuhiro Yoshikawa
Kyoto 1993

CONTENTS

PART IV. INFRASTRUCTURE FOR THE COSMO-CREATIVE SOCIETY

PART V. PLANNING AND POLICY PERSPECTIVES

CHAPTER 1

Logistical Dynamics, Creativity and Infrastructure

Åke E. Andersson, David F. Batten, Kiyoshi Kobayashi and Kazuhiro Yoshikawa

1.1 INTRODUCTION

Few would doubt that we are presently in the midst of an era of fundamental change. The powerful motor driving this latest revolutionary transformation of the post-industrial economies is generally thought to be technological. But such an explanation seems too narrow. The revolutionary age in which we now find ourselves is also a cosmo-creative one. Cosmo-creative activities, such as global research and development programs, are expanding as a result of the ongoing transition towards a knowledge-intensive society and some fundamental changes to the world's logistical networks.

Logistical networks are basic to all forms of movement and human interaction. Essentially, they are those systems in geographical space which facilitate the movement of commodities, people and money in association with the production or consumption of goods and services. Such networks also facilitate knowledge accumulation and diffusion. Furthermore, they foster face-to-face contacts which are essential for creative exchanges. The overall outcome is that we are entering an era in which creativity - both individual and collective - is gradually emerging as a vital component of dynamic comparative advantage on a global scale.

Logistical networks are observable and permanent (highly durable) systems of infrastructure which sustain all types of motion and interaction between global networks of cities. There is hardware and software. The hardware is made up of physical networks which promote interaction - such as conference buildings, computers, transportation and telecommunication systems. The software corresponds to those networks which embellish human infrastructure - such as education, the arts and science. Many of these softer networks are intangible to all except those intimately involved.

1.2 THE FOUR LOGISTICAL REVOLUTIONS

It is instructive to explore the catalytic role of logistical networks with the help of qualitative dynamics. What emerges are some historical and social preconditions which have partly shaped the development of each and every metropolis during the last millennium, as well as some new preconditions which will affect their future development in tomorrow's international economy. The well-recorded sequence of events which have transformed the world economy since the Carolingian era can most likely be explained by the changing structure of what we have broadly defined as logistical systems. In other words, the principal changes of population, production, trade, culture and institutions are triggered by slow but steady changes in the world's logistical networks.

In recent years, sudden and unexpected discontinuities have become central issues in political and planning debates about the future of metropolises. As discussed in the contribution to this volume by *Andersson*, it turns out to be useful to look back into history for evidence of such behaviour in the evolution of cities as a global family. One illuminating example is the path-breaking analysis by the mathematician, Alistair Mees (1975). He took the hypothesis of Henri Pirenne (1925) as the starting point for an analysis of sudden changes in the specialization pattern of a set of trading regions. Pirenne's hypothesis stated that the central cause of the revival of European cities and towns in the late middle ages was the emergence of free trade and consequent improvements to the transportation systems.

Andersson has since broadened the Pirenne-Mees analysis to a more general one, claiming that the sequence of fundamental changes in the world economy over the last millennium can be explained by the changing structure of logistical systems. In other words, the great structural changes of production, location, trade, culture and institutions have been triggered by slow but steady changes to the associated logistical networks. The development of cities and their inter-regional economic relations in the world from the years 1000 A.D. until 2000 A.D. may be perceived in terms of four logistical revolutions, namely the following:

(1) Emerging in Italy during the 11th century and ending in Northern Europe during the 16th century,

(2) Emerging in Spain, Portugal and Italy during the 16th century and ending in Northern Europe during the 19th century,

(3) Emerging in England during the 18th century and ending in the developing countries, probably during the 21st century,

(4) Emerging in e.g. Japan, USA, Switzerland, Germany and Sweden at the end of the 20th century.

It is very probable that we are now witnessing the emergence of the fourth logistical revolution associated with the growth of information processing and communication capacity as well as the growth of the knowledge base. This development goes hand in hand with further improvements to the transportation system, especially to the structure and operation of the air transportation networks. Such developments tend to increase the discrete network character of the global economy. Continuity of places is thus becoming less important.

A key characteristic of the fourth logistical revolution is the slow expansion of the knowledge base by virtue of creative minds. Knowledge-producing capacity expands steadily to a critical point whereupon a major structural change in the regional economy must eventuate. With this explosion of knowledge-intensive activities comes a shift of paradigm from the component and factory-oriented engineering of the third logistical revolution to a systems architecture and systems analysis perspective. In any industrial sector structured according to such a paradigm, *synergisors* (or non-separable dynamic effects) should not be seen as imperfections but instead as potential profits to be secured by the correct design of the production system as an holistic unit. A slow but steady reduction in the costs of information transfer, coordination and control over longer distances has been an important catalyst of the emergence of the systems paradigm.

Thus, the current logistical revolution may be interpreted as a synergistic interplay between various creative minds. Certain individuals and regions are already better-equipped to exploit the advantage of this revolution. A city can evolve into a creative region only if its creative environments develop intensely and extensively. The preconditions for creative activities are determined at the level of the individual person. Individual creativity is seen as a formulation of novel problems in interaction with our systems of perception, memory and motivation. The Chapter in this volume by *Smith* investigates an analytical foundation for this analysis at the level of the creative individual. Smith considers that one's inner world is shaped from within, by means of ultrashort processes rooted in private, often infantile experiences. Smith tries to answer the important question he raises in his Chapter: how willing is the receiver, in spite of the convincing illusion of an independent outside world, to admit its private origins, to reconstruct and use them? His experimental research shows that the more open to näive reconstructions the more creative is the individual. He also discusses aspects of creativity in relation to aggregation, self-image and aging.

Logistical revolutions seem to provide part of the necessary catalyst for creative expansion. If creativity is to proliferate within a region, additional conditions must also be satisfied. Some of these conditions are environmental ones, catalyzing interactions among creative individuals. The Chapter by *Haken* on 'Synergetics as a theory of creativity and its planning' discusses the basic principles of synergetic interactions for creativity. He claims that cities or whole areas can be conceived as a giant organism, with inputs and outputs of energy, matter and information. Haken explains some of the basic principles of self-organization of organisms, which have been unearthed by the interdisciplinary field of synergetics (Haken, 1983). He shows how these principles work and enable a city system to become creative; how qualitatively new features of a system may emerge through synergy. Haken points to the importance of coherence among the individual parts of a system and concludes his Chapter by stressing how coherence and emergence of new properties can be planned.

1.3 KNOWLEDGE CREATION AND KNOWLEDGE-HANDLING LABOUR FORCE

In a pioneering study of the U.S. economy, F. Machlup (1980) concluded that a group of industries which he collectively called 'the knowledge industry' already accounted for some 29 percent of the U.S. Gross National Product in 1958. Machlup also provided a new perspective on the labour force by revealing that 'knowledge-producing' occupations exceeded all other groups of occupations in terms if growth since the turn of the century.

Machlup's concept of knowledge-producing occupations was rather broad, combining job categories that create new knowledge with those that communicate existing knowledge and information to others. The group of communicators includes some who transfer knowledge directly, but also a host of technical and clerical workers who deal mainly with routine information. Including this large 'supporting cast' of information-handling employees tends to give a misleading picture of the true size of the 'knowledge-handling' work force. It is necessary to rework Machlup's figures in order to distinguish more clearly between knowledge-based and information-based occupations. For example, if we restrict our definition of knowledge-handlers to those formally engaged in education, R & D, journalism and artistic activity, the share of knowledge-handling jobs in the Swedish economy grew quite rapidly from 10 % to 18 % of the labour force between 1960 and 1980. Should such a trend persists to the turn of the millennium, the knowledge-handling share will rise to approximately 30 % of the Swedish labour force.

Similar changes have also occurred in Japan and other parts of Europe. If anything, the rates of expansion have been faster in Japan and Europe than in the United States during the seventies and eighties, as the size of the knowledge-handling labour force in the United States has leveled off. This mushrooming growth in knowledge-handling occupations has not been confined to knowledge-intensive sectors within manufacturing industry. Across almost every segment of manufacturing industry, a deepening of the knowledge content has occurred within the labour force. This may be illustrated from both a sectoral and an occupational viewpoint. Consider the sectoral development of employment in manufacturing industry within the metropolitan regions of Stockholm during the last two decades. The fact that the share of knowledge-handling occupations in total employment has expanded at rates varying between 1.8 percent and 3.2 percent confirms that the internal restructuring of Stockholm's manufacturing industry in favour of knowledge-handlers has been as fast as in the entire Stockholm economy.

The pace of change in the creative arena is rapid. Mushrooming growth among firms responsible for information-processing, telecommunications and air transport capacity is greatly facilitating point-to-point contacts between many dispersed locations, thus increasing the network character of the world economy. The education system is expanding the knowledge base of urban populations, thereby promoting more efficient use of computers, telecommunications and rapid transit systems. Knowledge about important innovations can spread rather quickly under these conditions. As these tendencies proliferate further, the geographic continuity of regions and the relative size of places in a local context may become less important than they were in the past. Instead we must concern ourselves with diverse changes permeating across interdependent networks. Furthermore, close links may be forged between places of complementary function rather than simply on the basis of distance or demand thresholds.

Because knowledge-intensive production units are rather sensitive to the availability of labour with high levels of education and experience, their relocation strategies tend to favour regions with universities and other advanced educational and scientific institutions. They also favour regions in which higher levels of scientific exchange take place. In their Chapter, *Anderstig and Hårsman* claim that the growing knowledge orientation has made availability of skilled labour and R & D resources a more decisive location factor. Their empirical study indicates a higher mobility among people with knowledge oriented occupations than among other occupations. They show that this tendency in Sweden fully replicates a US study, demonstrating that life-cycle characteristics have a dominating influence on the decision to migrate. They show that different regional attributes and amenities have a significant influence on the migration decision, and conclude that a region which is more attractive for people with high-tech skills will also be more attractive for high-tech firms.

1.4 LOGISTICAL NETWORKS AND CHAOTIC DYNAMICS

The present tidal wave of social and economic change associated with the fourth logistical revolution also has much in common with the unpredictable evolutionary path of science. In such a dynamic and nonlinear world, predictability becomes the exception rather than the rule. The *status quo* may be a society in which chaos, instability and volatility not only govern general patterns of behaviour, but are also key phenomena to be understood by the planning profession.

The classic theory of systems dynamics is unable to provide us with a convincing explanation of these phenomena. Thus we find ourselves in need of paradigm shifts in both the natural and the social sciences. Recent advances in nonlinear dynamics have cast grave doubts on our abilities to fully comprehend, plan or control social systems (in the classical sense of the words). Newly emerging theories may provide us with a deeper understanding of changing human behaviour and social, economic and institutional systems. They suggest that it is difficult to plan successfully for the future without incorporating sufficient resilience or robustness to accommodate unpredictable swings associated with logistical networks. Naturally the following question arises: Does unpredictability imply uncontrollability of socio-economic systems?

In China and Japan, the notion of unpredictability has been injected into daily modes of thinking in a rather natural manner. The ancient Chinese used to consult the 'I Ching' (or 'Book of Changes'), reputed to be the oldest book of mankind. The 'I Ching' tells us that while changes can sometimes (appear to) be horrendous, the wise person will be cautious, patient and perseverant. It does not prescribe any scientific methodology to adopt in the face of uncertainty, but rather it offers some guidelines and precautions which the modern decision-maker might also apply to situations of uncertainty. If the task of the wise person is to understand the great rhythms of the Universe and to work in harmony with them, then the 'I Ching' provides some qualitative guidelines which may prove more helpful than quantitative prescriptions.

Although its obscurity has caused many Western scholars to dismiss it as a collection of magic spells, the 'I Ching' philosophy conceptually has much in common with the modern theories of nonlinear dynamic systems. Conversely, it has little in common with the classical scientific approach to prediction and uncertainty. The classical school of logical positivism tends to the view that prediction is simply a synonym for scientific explanation. Any well-founded scientific theory which can explain past changes in a social system should be capable of predicting its future trajectory as well.

Aside from obvious difficulties in defining 'well-foundedness', the modern theories of dynamical systems have cast other doubts upon the validity of the classical scientific methods of prediction. They have demonstrated that even the very simplest of models can sometimes generate extremely complex and chaotic systems behaviour. This makes prediction almost impossible unless we have an extremely detailed and accurate description of the initial and permissible states of our system - including the precise values of all system parameters and the range of values which our key variables are able to assume. This calls for detailed observations far beyond the powers of the classical scientist.

At this juncture, the 'I Ching' enters again! Whereas Western science is largely based upon the principle of causality, the Chinese mind seems to be mostly preoccupied with the chance aspect of events. Far from being an axiomatic truth, causality passes almost unnoticed. As Jung puts it:

> 'While the western mind carefully sifts, weighs, selects, classifies, isolates, the Chinese picture of the moment encompasses everything down to the minutest nonsensical detail, because all of the ingredients make up the observed moment' (Preface to *I Ching* translated by Wilhelm, R.).

This is actually a rather practical stance to take. At best, causality can be regarded as a statistical truth and can never be absolute; it provides us with a working hypothesis of how events evolve from other events. The 'I Ching' principle of 'synchronicity' takes the coincidence of events in space and time as meaning something more than mere chance, namely, a peculiar interdependence of objective events among themselves as well as with the subjective (psychic) states of the observer(s). This attention to qualitative detail is paramount in a nonlinear dynamic world, since the unfolding trajectory is so sensitive to the initial state and to the way in which we specify the system itself.

Modern science may be moving in the direction of the 'I Ching'. The axioms of causality are being shaken to their very foundations. It now seems evident that a 'meeting of the minds' may occur eventually, as modern physics attempts to understand the role of the observer in experimental situations. Indeed chance is commonplace in nature, whereas a natural course of events conforming to specific laws is almost an exception. If we therefore decide to acknowledge that chance (i.e. unpredictability) plays an important role in social and economic change, we must begin to invent new methodologies which place complexity, nonlinearity, and instability at the core of the planning domain. This volume addresses such challenges in the Chapters of Part II by *Saari, Zhang* and *Kobayashi*.

In his Chapter, *Saari* summarizes the revolution in the mathematics and economics of chaos over the last decade. We now know that even simple, deterministic systems may not be as predictable and orderly as once thought. Instead, the theory of chaos teaches us that these systems can result in behaviour so unpredictable that it may appear to be random. In Saari's expository Chapter, some of the basic causes of 'chaos' are identified. Then, the implications of 'chaos' with respect to some of the basic themes of the book - including decision theory - are discussed.

Zhang's Chapter explores some aspects of the complexity of economic decisions in a dynamic world. Zhang is mainly concerned with the relationship between economic instability and anticipatory behaviour. He also explains some of the characteristics of entrepreneurship and examines the complexity of planning in mixed economies against the background of chaos and anticipation.

Kobayashi presents a new analytical framework for logistical network equilibria with incomplete information. In his model, the basic unpredictability of network users' behaviour stems from the differentiation of their private information. If users respond to different private information, their behaviour in each round looks chaotic to the observer. The purpose of his Chapter is to show that there emerges a stochastic network equilibria, if the fluctuations of private information are subject to a certain probabilistic distribution and users are rational in their choice. It is a reinterpretation of Harsanyi's Bayesian Nash equilibria, but differs from Harsanyi's in the explicit consideration of users' rational expectations formation.

1.5 COSMO-CREATIVITY

Science does not advance along a straight line. The collective advances of science as a whole show a definite alternation between relatively brief eruptions which lead to the conquest of new frontiers, and long periods of consolidation. As Koestler (1964) has stated:

> 'Mental evolution is a continuation of biological evolution, and in various respects resembles its crooked ways. Evolution is known

to be a wasteful, fumbling process characterized by sudden mutations of unknown cause, by the slow grinding of selection, and by the dead-ends of over-specialization and loss of adaptability ······ Moreover, there occur in biological evolution periods of crisis and transition when there is a rapid, almost explosive, branching out in all directions, often resulting in a radical change in the dominant trend of development'. (p.226)

Thus, science evolves in an unpredictable way. Far from being smooth and certain, its path of evolution is jerky and unpredictable. Part of the explanation for this is that many significant advances in the history of scientific thought have occurred by unexpected mental cross-fertilization between different disciplines. If such bisociations are the essence of creative activity, then it is important to recognize both the local and cosmopolitan components of knowledge. In other words, are there recognizable limits to the formalization of (shared or common) scientific knowledge attributable to the semantics entailed in different cultures (local knowledge)? Once we enter this humanistic arena, the ensuing discussion takes on a qualitative character.

As Robert Rosen (1991) has pointed out, there is a duality between the qualitative and the quantitative. In the sciences, this dichotomy rests on presuppositions concerning the nature of material reality and on how we obtain knowledge about it. These scientific presuppositions also have mathematical counterparts, where the duality lies between syntactics and semantics and is centered around the notion of formalization. The advantage of taking this approach is that there exists a theorem which actually helps to partly resolve the issue, namely Gödel's Theorem. Science is concerned with systems of causal entailment in the phenomenal world. Gödel's Theorem asserts that a truly complex system is more complex than any formalization in which the entailment is purely syntactic (Rosen, 1991). There are thus qualities pertaining to that complex system which will remain unencoded by any such attempt. In the case of knowledge, this is simply illustrated by the distinction (referred to above) between local and common knowledge. Local knowledge creates the semantics which adds complexity to any simpler syntactic entailment. Thus the duality between the quantitative and the qualitative boils down to a relative question of simplicity versus complexity. Since local knowledge is very slow to change and heavily dependent on the local environment, diversity is a natural outcome of knowledge exchanges on a global basis. Alternative formalizations are always possible and external observers may actually be responsible for deciding upon the particular formalization to be adopted.

The above discussion serves as a stepping-stone towards a definition of cosmo-creativity. A complex system like the creative individual seemingly defines formalization. Part of the difficulty stems from the fact that this complexity is more of a subjective property of an isolated system. It can only become more objective when our formalization takes into account the society or environment with which the creative individual interacts. We have chosen to call this broader creative system, in which many creative individuals interact with one another, the *cosmo-creative* society. On the one hand, it has a cosmopolitan dimension because creative exchanges permeate cultural borders and take on an international character via the many programs of scientific cooperation and exchange. On the other hand, it has a local dimension which is intimately associated with the local environment. The complex interplay between these two facets of knowledge exchange are at the foundation of the cosmo-creative process.

The evolutionary path of a cosmo-creative society has much in common with that of science as a whole. Being largely qualitative, it is mostly unpredictable. For example, the cosmo-creative city must be flexible enough to promote interdisciplinary contacts across different cultures but, at the same time, it must be sufficiently robust to exploit controversial discoveries and to withstand unexpected changes of direction. It is an extremely dynamic world, one which twists and turns in different directions as creative whims may dictate. To fully comprehend the cosmo-creative society requires a deeper understanding of the catalytic role of key individuals involved in logistical networks and the nonlinear character of the dynamic process of networking. Many of the Chapters contained in this volume are motivated by this important new perspective. In particular, the Chapters of Part III by *Okada, Batten* and *Johansson* highlight the above mentioned issues relating to the cosmo-creative evolution of logistical networks.

In his Chapter, *Okada* claims that in the context of the fourth logistical revolution, entrepreneurship should be seen from the viewpoint of multiple agents forming a network. He then seeks to model the mechanisms related to the development of 'cosmo-creative entrepreneurship', in the form of a communication game involving the interplay of multiple players. He concludes that though a local region is handicapped when networking with principal agents, it could possibly overcome these problems if it becomes open and flexible enough to the external world, assuming that the innovation is a vital one.

The natural consequences of cosmo-creative dynamics are that future phases of growth and decline among systems of cities may be intimately associated with each city's relative position in various cosmo-creative networks. It also suggests that competitive leadership may not be restricted to larger urban centers which have served traditionally as the seedbeds of creative activity. Smaller settlements may prosper by establishing international cooperation in creative fields such as science, literature and the arts. Thus knowledge-sharing becomes a more realizable process and the new cosmo-creative era gathers momentum. In his contribution to this volume, *Batten* discusses some of the important mechanisms behind the dynamics of growth and decline among systems of cities. His Chapter highlights findings from some recent studies concerning metropolitan development in the industrialized world. Two particular aspects are stressed (1) a need to expand our domain of analysis for city systems-from that of centers within subnational hierarchies of settlements to that of cosmopolitan hubs, nodes, corridors in a complex system of independent global networks; and (2) a need to revise our traditional theories of the urban development in order to cater for the synergies inherent in these networks, and the fact that collective stocks of knowledge are a function of the complete constellation of knowledge-enhancing cities and do not reside in any single center alone.

Johansson examines infrastructure in the form of the cosmo-creative built environment and associated networks which facilitate interaction between individuals and between organizations. Johansson demonstrates that when an urban system is characterized by general infrastructure, then the self-organizing mechanisms for the entry and exit of activities play a crucial role for creative evolution. His Chapter also recognizes that infrastructure capital ages in a technological sense over time. A model is used to explain mechanisms which generate a cyclic process in which the infrastructure of urban zones switch between long intervals of investments and renewal, followed by similarly long intervals of aging.

1.6 INFRASTRUCTURE FOR THE COSMO-CREATIVE SOCIETY

It is now time to return to the question posed earlier: Does unpredictability imply uncontrollability of socio-economic systems? Not necessarily. One promising and plausible way to delimit chaotic phenomena is to create 'buffer' systems which can absorb or attenuate unpredictable fluctuations in the original system. Examples of such buffer systems abound throughout history. The creation of modern enterprises was an innovation which overcame many of the instabilities inherent in the capitalist system. Security markets have helped considerably to absorb price volatility among stocks and shares. Technology itself provides many solutions which diminish uncertainties and stabilize outcomes. In the midst of a chaotic world, new planning strategies should be sought that facilitate qualitative improvements to our existing societies and their institutions.

One of the key buffer systems which can assist in this regard is infrastructure. Tangible systems of infrastructure - like transport and communication systems - are not merely capital instruments which assist in the production process. Being durable, they provide considerable stability and certainty in the longer term. Being public in character, they also serve as a rather flexible 'arena' on which many different 'games' can be played. These two important attributes suggest that infrastructure can be conceived as something of a relatively slowly changing and highly public nature. It tends to retain its overall (macro) structure for many decades even though parts of it may undergo significant change more quickly. Relatively slow changes are orders-of-magnitude slower than the faster pace of changing activity patterns which might utilize the infrastructure.

The benefits bestowed upon society by infrastructure are often underestimated. This is because the true (long-term) benefits of infrastructure investments are impossible to measure within any short-term framework. The ultimate return to any society is a non-marginal one, since better or different infrastructure may create an entirely new stimulus for further investment by providing new opportunities which have never been foreseen. These non-marginal effects are true of both hardware and software. Qualitative improvements to human infrastructure, provided by the education system, may have significant costs in the short term but enormous advantages which accrue only over the longer term. Being qualitative in character, the long-term benefits of infrastructure are difficult to predict. They provide a substructure for many different activities. To a certain extent, they also establish some of the 'rules' or customs which must be observed in order to utilize their substructure efficiently. Rules of a collective character may open up new pathways of opportunity as well as imposing constraints (e.g. congestion). Since the outcomes are often unpredictable, the pertinent analytical tools are of a nonlinear character.

In the case of infrastructure, the design problem can be likened to the evolution of the billiard table. The precise form of this table has changed very slowly over many centuries. Being the exclusive property of the rich or royal, it was extremely expensive. As interest grew in the game itself, cheaper versions soon proliferated. Thus the billiard table made the transition from a purely private good to a club good. By introducing new games such as snooker and pool, public interest continued to grow. The club good was gradually changing to a public good. Nowadays, almost anyone can play for a small fee. New ball games may also be devised which make use of the same table. The rules of these games are delimited to a certain extent by the precise nature and form of the table. But within the limitations set by the table itself, the number of possible games is very large.

Of particular importance is the fact that a poorly constructed table will invariably produce a great amount of chaotic motion. The rebounding trajectories of the balls are extremely difficult to predict. To reduce uncertainty and improve the quality of play, the surface and the cushions must be crafted very skillfully. Thus the quality of the craftsmanship is of vital importance in exercising some control over the frequency of chaotic motion. Infrastructure 'craftsmanship' has much in common with the billiard table. It has evolved rather slowly and taken on more of a public goods character. By providing a substructure for many different activities, it exerts a major influence on a wide spectrum of social and economic opportunities. Sometimes it facilitates, sometimes it constrains. As such it has a fundamental role to play in strategic (long-term) planning. When viewed in terms of key service attributes, the evolutionary patterns of infrastructure become relatively clear and predictable. Both transport and communication systems have evolved by embracing faster, more flexible and reliable means of interaction offering a much larger choice of destination possibilities. Although these evolutionary characteristics are rather clear, the overall changes have been very slow to materialize and many of the opportunities provided by them are yet to be fully realized.

Part IV of this volume assembles some Chapters related to the infrastructure debate (in addition to those presented in Part III). Several authors discuss the roles of different types of infrastructure and their impacts on urban structure and regional economies. As *Andersson* defines 'infrastructure' earlier in this volume, the term covers the broad category of public phenomena which are highly durable. It can be tangible or intangible. Chapters by *Sasaki, Mun, Kobayashi et al.* and *Kashiwadani* discuss tangible types of infrastructure, whereas *Okumura et al.* investigate the role of intangible infrastructure like regional reputation.

Sasaki and *Mun*'s contributions to this volume both attempt to analyze the impacts of information technology on urban spatial structure. *Sasaki* investigates the hypothesis that firms tend to be dispersed and locate their two functions (administrative function and routine function) at different locations in an information society. This hypothesis is based on the conjecture that it will become more profitable for the routine-activity sector to be located in suburban areas, where both the land rent and wage rate are lower. This also implies that face-to-face communications and contacts made by the routine-activity sector may be substituted by communications using information technology. On the other hand, the Chapter by *Mun* complements *Sasaki*'s Chapter by focusing rather specifically on the impacts of information technology on office locations and the factor demand substitution between face-to-face and telecommunication contacts. Mun formulates an office location model endogenizing the demand generation for telecommunications and face-to-face contacts. He also investigates the short- and long-run effects of technical progress in telecommunication on travel demands and office locations. The Chapter by *Kobayashi, Sunao and Yoshikawa* contrasts the foregoing Chapters by *Sasaki* and by *Mun* in the sense that they focus on 'meeting-facilities' instead of link infrastructure (such as transportation and telecommunication facilities). Meeting-facilities form the nodal infrastructure which facilitates efficient knowledge exchange between creative individuals.

The population of most major cities in the developed world has not increased in recent years, but their land use patterns are changing intensively. We need new models to project housing stock patterns in such non-growth cities. In his Chapter in this volume, *Kashiwadani* presents a model which can estimate housing stocks by housing type, corresponding to the number of newly constructed houses. The model reveals an input output table of available housing vacancies. These types of models estimating infrastructure stocks are indispensable for the metabolic planning of urban systems. Kashiwadani concludes his Chapter by demonstrating the applicability of the model through an empirical study conducted for the Osaka Bay Area.

Okumura, Yoshikawa and Hideshima's Chapter on 'Optimal regional investment control using hallmark events' introduces a new type of regional infrastructure, i.e. regional reputation. Regional reputation is intangible but significant for the development of regional tourism. In the short run, a hallmark event is one vehicle to lift regional income; conditioned by the basic properties of the event itself as well as regional reputation. In the long run, it turns out to be an important determinant of regional reputation. The authors formulate an optimal control model for regional investment which addresses the timing of hallmark events and the investment intensity of infrastructure within the host region.

1.7 PLANNING AND POLICY IMPLICATIONS

During the sixties and early seventies, many urban planners and regional scientists tended to perceive the future of the metropolis simply as a problem of optimal layout and design. This common view was, of course, partly valid, but only within a very limited domain. Today's modern metropolis is rarely a new town. Surprisingly few of the world's major cities have actually evolved in accordance with a prescribed master plan. Most have developed organically over many centuries, experiencing frequent fluctuations in their economic property and significant changes to their cultural, social, technological and political conditions. In a system as vast and complex as a major metropolis, quantitative optimization is a microscopic perspective and the explanatory power of conventional multi-objective methods remains debatable.

The serene days of strong economic growth, smooth urbanization trends, and numerous opportunities for optimal planning and design of cities are rapidly becoming a distant memory. Economic sluggishness tempered with abrupt and sometimes unexpected changes appear to be today's routine. A growing recognition of such discontinuity or structural change prompts a fundamental need for new theoretical frameworks and analytical tools which can probe beyond the boundaries of the traditional introspective paradigms provided by optimization theory and the urban lifecycle model.

In the school of neoclassical economics, technologies are supposed to be chosen by the cost minimizing or the profit maximizing behaviour of individuals, firms and organizations. The underlying rationality of human behaviour is a very significant factor for understanding the repercussions of technological changes. By applying this assumption of rational progress, we can predict (to some degree) the future states of technologies. On the other hand, rationality does not play any decisive role in predicting the future states of human knowledge. There is no way to know today what we can only know tomorrow (Popper, 1957). Thus, in periods of innovation, global change, and creativity, we must expect the unexpected. The emergence of something unexpected can pave the way for new opportunities for science, business and arts.

In the face of growing uncertainty, planning strategies must become more flexible. As planners learn to understand the complex dynamic processes and interdependencies associated with the real world, they will also begin to appreciate the starategic importance of slowly changing systems such as logistical networks. Traditional planning paradigms have identified desirable states towards which decision makers have striven. The new planning approach may be identified more with abilities to understand how societies and their economies evolve over time. Rather than ensuring that decision makers' choices lead to their desired ends, the emphasis will be on whether their decisions lead to ends (as states) which are ethically desirable for present and future generations.

Our present understanding of society and human affairs is primitive and fragmentary. We know so little about some cultures and their complex institutions that any attempts to predict their future trajectories would be doomed to failure. In such circumstances, planning cannot hope to be comprehensive. Even the cherished *scientific* approach to planning cannot be supported, since the classical methods maintain that their scientific findings can unify nature and human societies. This notion of *comprehensiveness* underpins both the planning and scientific professions. Yet comprehensive prediction is clearly impossible if the system involved (e.g. an economy or a society), or just a small part of it, is susceptible to extremely complex and chaotic behaviour. We are now beginning to see the emergence of new perspectives and new paradigms underpinning human behaviour. They do not attempt to tie together all aspects of human behaviour in one grand framework. Rather they begin to search for key factors which are of strategic importance as the architectural foundations of societal change. Some of these factors have been discussed in this volume - slowly changing factors such as culture, knowledge and logistical networks. It is on these foundations that the new paradigms of the next millenium must be erected.

After reflecting on the cosmo-creative planning approaches described in this volume, *Yoshikawa* presents an ambitious vision for the Osaka Bay Area as a pioneering metropolitan arena for cosmo-creative renaissance. Instead of formulating a comprehensive set of planning goals and targets, he attempts to construct a conceptual platform which can underpin a great diversity of innovative efforts by both the public and private sectors who share a common purpose - to transform the Osaka Bay Area into a truly cosmo-creative metropolis. This conceptual platform consists of those classes of tangible and intangible infrastructures which are clearly indispensable for this new type of planning exercise. Yoshikawa also stress the importance of visions for the future. It is only by visions that a future cosmo-creative world can be visualized. Our modest hope is that this book may help to carry this message to the world of researchers and practitioners engaged in the planning professions.

REFERENCES

Andersson, Å.E., 1985, *Creativity: The Future of the Metropolis,* Prisma, Stockholm (in Swedish).

Haken, H., 1983, *Synergetics, An Introduction, 3rd ed.,* Springer-Verlag, Berlin, New York.

Koestler, A., 1964, *The Act of Creation*, Pan Books, New York.

Machlup, F., 1980, *Knowledge and Knowledge Production*, Princeton University Press, Princeton.

Mees, A., 1975, "The revival of cities in medieval Europe", *Regional Science and Urban Economics*, 5:403-425.

Pirenne, H., 1936, *Economic and Social History of Medieval Europe*, Routledge & Kegan Paul, London.

Popper, K., 1957, *The Poverty of Historicism*, Routledge & Kegan Paul, London.

Wilhelm, R. (translation), 1951, *I Ching (Book of Changes)*, Arkana.

Rosen, R., 1991, *Life Itself*, Columbia University Press, New York.

CHAPTER 2

Economic Structure of the 21st Century

Åke E. Andersson

2.1 DYNAMICS AND CHAOS

A number of mathematical studies have shown that predicting the future with the assistance of non-linear models can be an exceedingly risky task. Even dynamic simulation models with a not too complex structure can give rise to unpredictability, meaning that even a very small change of some parameter or initial value can lead to a completely different prediction. Non-linearity breeds chaos.

These essentially mathematical studies of dynamic systems have recently been supported by applied mathematical studies of economic systems, showing that chaos can occur in not too unrealistic mathematical models of the business cycle (Puu, 1989). Combining the multiplier principle with a non-linear accelerator determining investments, it can be shown that for reasonable ranges of the model, parameters can be caught in chaos with or without growth. If this is the case, short term economic forecasting with non-linear econometric models is no longer a scientific undertaking. The outcome would be as reliable as clairvoyance or fortune telling.

However, this conclusion does not imply that long term analysis of the structural evolution of an economy is useless. Medium term meteorological forecasting by computer methods has turned out to be highly unreliable. This has not prohibited the use of long term evolutionary models of the climate. We have to make the same distinction between short and medium term econometric forecasting and long term evolutionary analysis based on their application to mathematical studies of dynamic systems. The analysis of this Chapter falls into the latter category of dynamic economic studies.

2.2 FAST AND SLOW PROCESSES

von Neumannn (1936) is said to have observed that it would be impossible to live in a world of non-linearities if everything proceeded on the same time scale. If we, for instance, could imagine such a world, it would mean that anyone entering a house would displace the positions of the walls, windows, ceilings and floors. With many people moving in the rooms and with a sufficiently strong impact on the building from their moves, it would be impossible to interact safely in such a building. Dancing or even walking slowly could be dangerous to your health. Fortunately, the world we live in does not have these dangerous properties. Most of the time the houses we live in, the roads we drive on and the organizations we work in are much more stable and stabilizing than the humans acting on or within them.

This difference in stability is a consequence of the differing time scales relevant to different dynamic processes. In the analysis of communication and transportation flows, the relevant time scales range from seconds to a few weeks. In these cases a complete dynamic analysis is often not needed. The calculation of a general static equilibrium is often sufficient.

Other economically important entities, such as railroad and road networks, require decades to centuries for a proper analysis of their evolution from one equilibrium to another. At any point in time a road network might look as if it were in a completely stable equilibrium. At any instance people can use it safely as a stable equilibrium condition of their daily activities. And yet, there is a slow and steady deterioration of any road network, threatening its long run viability. Similarly, bottleneck problems trigger extensions and changes of the capacity and topological structure of the network. However, the approximate stability of the arena, generated by the slow processes, is enough for the establishment of a true equilibrium of the economic, social, and cultural games to be played.

2.3 PRIVATE AND PUBLIC PHENOMENA

Some economic products are of a purely private character. The consumption of food is one example. When somebody eats a banana or a piece of candy it can be regarded as a purely individual or private phenomenon in the sense that the consumer is the only one who enjoys the product; i.e. the only one whose utility is influenced by the consumption, while the commodity is then completely excluded from use by somebody else. However, if somebody listens to a compact disc recording, the situation is quite different. Confined within a home, others close enough to the record player will also benefit or be annoyed by the sound, while simultaneously consuming the services of the record and the record player. This means that in this case we can no longer consider the phenomenon as being purely private or individual. It is a case of a public phenomenon. And in fact most commodities are in this sense public. But most of them are locally public only. By forming a small association of individuals, e.g. a family or a club, such local public goods can be treated as if they were private. Parents might be enslaved by the musical taste and record playing of their children, but these matters are normally resolved by some negotiation game within the family. The situation would be quite different if the record was played by a disc jockey in command of a national radio communication network. In this case there would be a steady information flow to be absorbed by anyone who had a receiver at their disposal. And the addition of another ten thousand receivers does not limit the usage by others. Similarly, if somebody is annoyed by the music they can turn off their radio. This is a case of a pure public good or a pure public phenomenon.

Many communication networks do have this property. However, in many cases there are objective qualitative differences in the enjoyment of public goods. Living behind a mountain can decrease the reception quality from radio communication networks. And with transportation networks other qualitative - or more generally - accessibility - problems occur. Some of these accessibility problems are subjected to spatial regularities. Somebody living in the center of a transportation or communication network will obtain, on the average, greater benefits from the network than somebody living in the periphery of the network. Recently there has been a somewhat heated discussion about whether congestion on a network transforms the network from a public good into a private good. This conclusion is inconsistent with the classical definition of public phenomena. The fact that the accessibility in a network decreases for some of its users entails a reduction of the overall quality of the network and not a conversion of a public phenomenon into a phenomenon of a private nature. Any congestion can conveniently be represented by a somewhat more complicated accessibility function. In fact, congestion is always a public phenomenon in the sense that it is an interaction situation, influencing every one involved in the congestion situation. Congestion is a typical measure of the quality of a network. In this sense, quality and accessibility are two natural measures of the performance of public networks. But the two measures are not mutually exclusive. Sometimes, different measures of quality can be embedded within an accessibility measure, influencing individual or public decision making about the usage of the network.

2.4 INFRASTRUCTURE AS SLOW AND PUBLIC

Some networks are very slow in terms of the time it would take to recreate the network completely. Recreating the European road network might require centuries of planning, negotiations and construction work. Setting up a new air transport network is a different story. This can be done within decades, and a new radio network can often be constructed within a few years.

Bearing this in mind, and the earlier observation that all networks are public phenomena, I propose the following definition of infrastructure.

> Definition: *Infrastructure is any public phenomenon which is highly durable, i.e. being subject to change on a relatively slow time scale.*

Although "relatively slow' might sound vague, it is in fact a rather precise notion. Relatively slow means slower by order of magnitude than the market processes regulating the exchange of market products, in the determination of a sectorial spatial structure of production, consumption, transportation and communication.

A slow and steady change must, according to this type of dynamic analysis, eventually lead to a rapid structural transformation at some instance of time, when the given equilibrium structure cannot be adapted to the changing arena. A sufficientlv large cumulated change of the logistical networks must thus sooner or later lead to a topologically different economic structure. The continuous evolution will eventually be disrupted by a structural transformation.

In the history of the world economy there have been a number of such logistical revolutions. Before the 12th century these transformations tended to follow a relaxation oscillation pattern. The most obvious examples stem from the growth and collapse of trading empires in the Mediterranean from the Fenicians to the collapse of the Roman Empire. The following cases provide one pattern of these logistical revolutions. The Fenicians, the Greeks, and the Romans slowly but steadily expanded their war machinery and trading capacity by opening up trading routes and new local, colonial regional markets for trade in commodities and slaves. Using some of the surplus for further extension of the controlled area, trade and wealth could be expanded, especially at the center of each new "world economy'. But the extension of each one of these world economies always tended to become too geographically large to permit long term sustainability. The costs of upkeep, repairs and other kinds of reinvestments in the extensive network sooner or later tended to outgrow the economic advantages of the extensions. The logistical network eventually became increasingly brittle, thus leading to increasing risks of the collapse of links critical to the stability of the whole trading system.

There seems to have been no way out of this problem. The more successful a colonial system became, the more threats to its internal stability would accumulate. If, for example, the central power of the network tried to stabilize the system by sharing the profits of trade with the colonies, the possibilities for these colonies to accumulate the strength necessary for an uprising would become greater. On the other hand, the more the central power exploited the colonies, the more reasons for an uprising accumulated. The more the trading empire cared for the upkeep and reinvestments in the growing network infrastructure, the fewer immediate advantages would be recorded by its own population. Any of these early trading empires was thus subject to a number of endogeneously growing threats to its viability and structural stability. This factor is consistent with our analysis for the rise and fall of Mediterranean empires preceeding modern times.

From the 11th century logistical revolutions lost this relaxation oscillation character. From then onwards, it is more reasonable to see the changes in a multi-dimensional consistently expanding mode of dynamics. And from the 11th century we can distinguish four different but essentially logistical revolutions leading to an increasingly interconnected world economy. Each one of these logistical revolutions follows a similar dynamic causal pattern: Firstly, the conditions of flows of information, knowledge, commodities and people along links are slowly but steadily improved. Some of these links are of a more public nature than others. In this sense they are critical and their inclusion into the network will thus have a systematically organizational effect; thus they are candidates in the explanation of major structural changes.

2.5 THE LOGIC OF STRUCTURAL TRANSFORMATION

To many economists (e.g. Brody, 1970) the study of economic structure is an issue of proportions within some production, capital or employment vector. Graphs showing the relative share of agriculture, manufacturing industry and services are typical of these approaches to structural analysis. For some purposes such a vectorial representation of structural change can be a suitable introduction to the issue but it can never become the focus of the analysis. From a topological point of view, any vector of positive values is structurally equivalent to any other vector containing positive values only, because any graph representing such distributions can be transformed into any other such graph by stretching and compressing without tearing. However, if one of the variable proportions should take on the value zero or suddenly increase from zero to a positive value, then the structural transformation would represent a case of a topologically meaningful structural change. Thus, the birth and the collapse of an industry in terms of production, capital or employment is a case of structural transformation within the "proportions school' of structural analysis. In short, the vectorial or proportions analysis of a structure of an economy is a very meager approach to structural economics.

By employing a network representation of an economy (or the equivalent matrix representation) of interactions, economic structure can be studied in a richer way. Studies of economic interactions are thus a natural starting point for any study of the structural evolution of an economy. Among the analysts and historians sharing this view of structural evolution we find A. Smith (1776), J. von Thünen (1826, new ed. 1966), J. Schumpeter (1934), W. Leontief (1949), O. Williamson (1986), M. Beckmann (1989), M. Morishima (1969) and T. Puu (1989). To these essentially theoretical analysts we ought to add a number of empirical economists and economic historians with a similar attitude to the structural problem, namely E. F. Heckscher (1935), H. Pirenne (1936), and F. Braudel (1979). In the core of their studies is a search for an understanding of the birth and death of employment and production in firms or industries, as well as the pattern of interactions in terms of flows of information or commodities in some organizational or geographical space. Concepts like interaction, tessellation, hierarchy, patterns and transformation are key words in such studies in addition to the classical concepts of equilibrium and stability.

It must be stressed that these authors share a common view about the important issues in the study of economic and social structure, rather than having any communality in terms of methodology or analytic skills. It is a conceptual communality rather than an analytical one. Within the framework of this conceptual attitude the remainder of this Chapter will be focused on the economic structure of market economies.

2.6 EQUILIBRIUM PATTERNS AND STRUCTURAL TRANSFORMATION

Any qualitatively meaningful study of economic structure must by necessity be dynamic and cover states of equilibrium, dis-equilibrium, and transformation from one structure to some other non-equivalent structure. The transformation problem cannot be relegated to some other modelling effort as it was by John Hicks (1965) in his otherwise important contribution *Capital and Growth.*

	Speed of Change	
Effects	Fast	Slow
Private (Individual)	Market Commodities *Games*	Education
Public (Collective)	Information	Networks *Arena*

Figure 2.1 : The arena and the games of policy making

Much of the dynamic analysis which facilitates the study of equilibrium, dis-equilibrium as well as transformations between topologically different structures is, in a deeper sense, hard to handle. The analyst cannot rely on properties that are easy to interpret and he must become acquainted with concepts such as bifurcation, turbulence, and chaos. There is also a great need for *empirical skills*, if unpredictability is to be avoided.

In the natural sciences many of these skills have been demonstrated by H. Haken (1983). Recently, the principles of synergetics as developed by Haken have been applied to structural analysis by Zhang (1991). Most of these methods are based on *dynamic* and *interactional* decomposition. (See also Chow and Hale, 1982; and Wiggins, 1988).

The way out of chaos in the modelling of evolution is often to investigate the empirical possibilities of such decomposition. If it is empirically true that some of the processes are distinctly slower than others - e.g. operating on a different time scale - and if it is furthermore true that some of these slowly operating processes concern variables of a public nature, then it is possible to end up with an evolutionary model which is predictable, while it is also able to handle equilibrium, dis-equilibrium and transformation analyses in an ordered and thus meaningful way. In practice, this means that if we separate material and non-material (or tangible and non-tangible) infrastructure from other variables and treat these infrastructure variables as the arena constraining the games to be played, then we will have a fair chance of understanding and even qualitatively predicting structural evolution.

Any dynamic modelling based on this principle will provide the possibility for analysing the existence and stability of an equilibrium structure of the markets, the political games and even the social structure as soon as the (assumed) almost given arena of infrastructure has been sufficiently specified. It is obvious that the more specified the arena, the less unpredictability there will be in the dynamic movements of the variables of the games.

2.7 LOGISTICAL REVOLUTIONS

In standard economic theory the starting point of the analysis is preferences and production technology. Demand functions are derived from the preferences given at the outset and the commodity supply is determined by the production technology available. If transportation flows are handled they play a secondary role in the links between firms and consumers. Conventionally, transportation and location issues are either outside the scope of the analysis or solved in a secondary fashion.

Such procedures are not consistent with the analysis of this Chapter. Material infrastructure in the form of logistical networks must then take on a similar role as the one played by the technology of production.

2.8 TRANSFORMATION FROM INDUSTRIAL SOCIETY TO C-SOCIETY

The industrial society can in all structurally important aspects be deduced from the infrastructure available for logistics and production, which can be seen as an important part of the logistical process. The infrastructural arena available in the 19th century was:

(1) Sea ports and steam ship technology available for long distance intercontinental transports. The North Atlantic basin constituted the core of this infrastructure.

(2) A railroad transportation network covering large parts of the east coast of the US and the Western parts of Europe, with connections in the US to the states around the great lakes and to certain other states. The major railroad system hubs were located in the metropolitan manufacturing cities, equipped with modern ocean ports.

(3) An average level of schooling of three years per capita of the employed population.

(4) A public knowledge infrastructure in the form of a scientific paradigm of physics and chemistry of a neo-classical type. This provided the foundation of mechanical engineering sciences and practice.

(5) A fairly general acceptance of the principles of private property rights, free entry and forced exits according to profitability or other efficiency criteria.

The first two conditions determined the focus on economies of scale through the production of homogeneous products, produced in large factories, located in the manufacturing metropolis close to the well equipped sea ports. The transportation technology simply required the use of economies of scale in production and homogeneity of shipments and production.

Another structural property was the vertical integration from the natural resource base through intermediary products to the final product to be distributed from the international ports. These vertical integrates followed the spatial structure of the railroad network and the waterways connecting the metropolitan ports and railway nodes with the natural resources needed for the outputs of the integrate.

Hierarchy in control, information and the diffusion of knowledge is another typical aspect of the industrial society. This was partly caused by the hierarchical structure of the networks used for communication and partly by the very skewed distribution of knowledge and the low average level of education. This characteristic distribution of knowledge was handled efficiently by hierarchical organizations and in the organization of defence and other public administration matters. Material and non-material infrastructure as well as historical traditions all contributed to the hierarchical organization of the administration of the industrial society.

The industrial society was, in comparison with the earlier agricultural society, extremely profitable and efficient. The combination of profitability and efficiency provided the necessary means for economic expansion of the industrial societies. It also provided funds for the improvement of the infrastructure. And with the improvement of the infrastructure followed an extended period of slow but steady creative destruction of the foundations of the industrial society itself. Roads were built. Roads which did not suffer from the indivisibilities and other economies of scale so typical of the railroad network. Telecommunication networks were created. Again, the creation of the telecommunication networks did not suffer from the same problems of scale economies that were typical of earlier combined transportation and communication networks. Education reforms were implemented and the new systems of education were both popular in scope and expansionist in terms of the number of school years. Finally, homogeneity of the sciences and of engineering was supplanted by increasing diversification, especially of physics. The long run consequence was a declining relative role of the mechanical paradigm. An increasing product diversification was supported by an increasing diversity of science and engineering ideas.

The creative destruction of the industrial arena simultaneously led to the creation of a new infrastructural arena requiring a transformation into a new economic and social structure.

2.9 THE EMERGING C-SOCIETY

The slow but steady destruction of the industrial arena in favour of the C-arena contains the following components: While seaports and shipping technology remain the basis of the long distance intercontinental transportation of commodities, the speed of such transport limits the use of that old network for the transportation of people and thus knowledge or information. The airline system and information and telecommunication networks are of steadily growing importance for long distance intercontinental transportation of people and information. The North Atlantic basin still constitutes one of the cores of this intercontinental infrastructure, but the Northern Pacific is a core of growing importance.

The sparse railroad transportation network of the industrial society has been supplemented by a much more dense road network covering not only coastal but also intracontinental areas. Besides this highly general network, there has been an ongoing establishment of energy grids like the European electricity and gas networks as well as other specialized commodity transportation networks. The implication of this process of investments into new networks for transportation and communication is the establishment of a multi-layered grid of networking opportunities, topologically completely different from the network structure of the industrial arena. There is no longer any great advantage of proximity to metropolitan manufacturing cities, equipped with modern ocean ports and major railroad connections. Intracontinental locations like Munich or Stuttgart in Western Europe, Kyoto in Japan, or Pittsburgh in the USA, are often more attractive alternatives to ocean-side locations.

The average level of schooling has increased from three years per capita of the employed population to levels ranging from ten to twelve years average schooling. A more important change is, however, the decreasing inequality of schooling. Most forms of higher education already existed in the early stages of industrialism, yet a significant share of the people were still illiterate. In most of the OECD countries there has been a consistent decrease of such inequalities.

The public knowledge infrastructure in the form of scientific paradigms of physics and chemistry has changed. With the development of quantum theory, molecular biology, information and computation sciences, the cognitive sciences, and new theories of dynamic processes, there is a large set of new technological and social potentials to be released. There is an increasing acceptance of non-national, interregional institutional arrangements in economic, social and cultural institutions.

The new multi-layered transportation and communication grid structure has not only given rise to new opportunities for intra-continental expansion. It has also opened up possibilities for a restructuring of production, distribution and industrial organization. One of the most important consequences of this seems to be the simultaneous expansion of the size of corporations and the reduction (at least relatively) of the size of production and distribution units within each one of these corporations. This simultaneous expansion and contraction of size is a clear consequence of the new dense network infrastructure. The dense structure of networks with their frequently thin links, connecting a large number of nodes, provides opportunities for small scale production and distribution units to exploit even small pockets of resources and demand. The total capacity of interaction within the multi-layered grid of transportation and communication links, however, at the same time provides great potential for cost sharing in matters of collective importance to corporations. Research and development costs can be shared world-wide as well as the costs of design, marketing and financial operations. It is economically efficient to develop economies of scale in the collectively important matters, while exploiting economies of scope by small scale diversification in the production and distribution of goods and associated services or services and associated goods.

Another important emerging characteristic of the C-society is the degree to which industrial organization and the associated control of production logistics is run by multi-level strict hierarchies. With the increasing number of spatially separated production and distribution units within corporations, hierarchical control becomes increasingly time and resource consuming. A corporation with a thousand production and distribution units can easily end up with hierarchical structures of control requiring hundreds of thousands of communication links. Self-organization of such production and distribution interaction is thus, from a purely efficiency oriented point of view, of great interest to the C-corporations. This tendency is further reinforced by the increasing average level and decreasing mal-distribution of formal education. No great risks are involved in the decentralization of decision making if most of the employees are well equipped with education and means of communication. But no corporation can survive without a common set of values and goals. The disintegration of the hierarchies thus results in an increasing importance of common value structures. Developing and diffusing a corporate ethic is probably an archetypical feature of the C-corporation.

Probably the most important characteristic of the 21st century economic structure is the C-region. With the increasing reliance on creativity and other cognitive capacities in the modern globally operating corporation, new locational criteria take on importance. The exploitation of natural resources and trading of rather simple commodities on the railroad and sea networks were the primary characteristics on the old arena. The modern corporation is much less oriented to the use of natural resources and much more oriented to profitability by increasing the complexity of commodities supplied to the world market. In terms of complexity, a modern car is qualitatively completely different from even the more advanced cars of the 1930's. And compared to their mechanical counterparts even a small computer is at other orders of magnitudes in terms of complexity.

The development of complex products requires skills, and skills are only to be had in regions equipped with higher education and the possibility of the synergy of such education by rapid and economically efficient modes of transportation and communication. In recent decades growth has been refocussed from many of the manufacturing cities along the ocean coasts to often much smaller regions, equipped with universities, institutes of technology and in prime locations with respect to international airports, highways, and fiber optical telecommunications. Looking into the future, the coming C-region of long-term viability will have a teleport, international universities, an international airport, and efficient highway connections with other national C-regions. Because of the increasing mobility of research personnel one aspect of long-term viability will also be the development of the cultural and natural qualities of life that can be realized in the C-regions. It must be stressed that we should not expect the C-regions of the 21st century necessarily to be of the same size as the megalopolis of the 20th century industrial society. Recent experience in the growth of regions in Europe and the United States would rather suggest that medium sized regions connected into C-corridors are more probable candidates for long-term viability as C-regions of the future.

REFERENCES

Beckmann, M., 1989, *Tinbergen Lectures in Organization Theory,* Springer-Verlag, Berlin.

Braudel, F., 1979, *Civilization and Capitalism, 15-18th Century,* Harper and Row, New York.

Brody, A., 1970, *Proportions, Prices and Planning, A Mathematical Restatement of the Labour Theory of Value,* North-Holland, Amsterdam.

Chow, S.N. and J.K. Hale, 1982, *Methods of Bifurcation Theory,* Springer-Verlag, Berlin.

Haken, H., 1983, *Advanced Synergetics,* Springer-Verlag, Berlin.

Heckscher, E.F., 1935, *Sveriges Ekonomiska Historia,* Stockholm.

Hicks, J., 1965, *Capital and Growth,* Oxford University Press, London.

Leontief, W., 1949, *The Structure of American Economy, 1919-1939,* Oxford University Press, London.

Morishima, M., 1969, *Theory of Economic Growth,* Clarendon Press, Oxford.

von Neumann, J., 1936, "A Model of General Economic Equilibrium (English translation from the German original)", *Review of Economic Studies,* 33:1-9:1945-6.

Pirenne, H., 1936, *Economic and Social History of Medieval Europe,* Routledge & Kegan Paul, London.

Puu, T., 1989, *Nonlinear Economic Dynamics,* Springer-Verlag, Berlin.

Schumpeter, J., 1934, *Theory of Economic Development,* Harvard University Press, Cambridge, Mass.

Smith, A., 1776, *An Inquiry into the Nature and Causes of the Wealth of Nations,* University of Chicago Press, Chicago, 1976.

von Thünen, J., 1826, *Der Isolierte Staat in Beziehung auf National-Ökonomie und Landwirtschaft,* Stuttgart 1966.

Wiggins, S., 1988, *Global Bifurcations and Chaos - Analytical Methods,* Springer-Verlag, Berlin.

Williamson, O., 1986, *Economic Organization: Firms, Markets and Policy Control,* Harvester Wheatsheat, New York.

Zhang, W.B., 1991, *Synergetic Economics,* Springer-Verlag, Berlin.

CHAPTER 3

The Creative Person

Gudmund J.W. Smith

3.1 INTRODUCTION

The term creativity, as used in psychology, has become chic. Many trend-setting terms live a short and hectic life and are soon forced to make room for new devices in the public show-windows. Not long ago a prominent creativity researcher asked if her beloved subject might also go out of fashion one day soon. But behind the term creativity one encounters an abyss of questions that ensnares us, at the same time tempting and frightening. The creative person, with whom we would like to identify, is commonly regarded as being filled to the brim of his mind with new ideas and projects: he views life from strange belvederes, formulates problems contrary to traditional beliefs, and turns accepted and seemingly self-evident conceptions upside-down. He may seem stimulating to those who work with him, an indispensable asset, but also exasperating, ego-centric, unreliable, and inconvenient. We usually accept a creative attitude as a constituent part of literary and artistic work, research, and technical construction. But creativity permeates all aspects of human activity, including our everyday business and communication.

In evaluations of the reasons for successful or failed careers, creativity appears to have supplanted intelligence, in the conventional meaning of the term, as the main explanatory factor. High intelligence was long believed to guarantee eminence within a number of fields, even practical ones. Occupational psychologists often managed to make reasonably reliable predictions of how young people would perform when studying various trades and professions, particularly if the assessment was profiled with regard to more specific intellectual components. To be sure, many highly intelligent people have had remarkable careers. At the same time, however, high intelligence does not guarantee that an individual will leave permanent traces in his craft such as, for example, ideological reorientations, inventions, or organizational change. His contributions could simply be the result of a first-class administrative job in a labyrinthine area, but wholly along traditional lines. Intelligence alone reinforces but does not fertilize.

The French psychologist Alfred Binet, the father of intelligence testing, seemed to have realized long ago that intelligence test scores were of limited predictive value. To balance this imperfection he coined the term 'divergent thinking' as opposed to 'convergent' thinking, the former breaking with, the latter conforming to established rules. The conventional intelligence test, moreover, measures level of achievement and does not reveal how the individual attained this level and the strategies used to solve the test problems. Among two test persons, one perhaps immediately grasped the essence of the problem, spotting the correct answer almost at once, while the other toiled along tortuous paths of reasoning to reach the same solution. These two very different minds would have been placed in the same category by a psychologist using a traditional test because they reached the same final result. Furthermore, in conventional intelligence assessment, the test person was never asked to formulate problems of his own; his role in the test was rather to find a suitable key fitting an old lock. People acquainted with locks beforehand were favored and performed well. It was advantageous for the final result if the person did not speculate too much about the deeper meaning of the problems presented.

The systematic study of creativity has proceeded along many lines of which two would seem to be the most important ones. Followers of Binet, like J.P. Guilford and his associates, devoted their efforts to the analysis of cognitive functioning, i.e., thinking in the broad meaning of the term. In their view, creative people ought to find unusual or 'divergent' applications for ordinary objects, like a brick or a newspaper, more easily than uncreative ones. A usual application of bricks would, of course, be house construction; a more unusual one, grinding the brick to mix it with solvent as paint pigment. The results of these tests correlated positively with other signs of creative functioning, but not impressively so. One reason could be that these researchers neglected the personality side of creativity, another that, being true Americans, they concentrated too much on worldly success, on fame, when choosing criteria to check their test instruments.

Clinical psychologists and psychiatrists, often with a psychoanalytic leaning, concentrated more on the creative person than on his cognitive functioning. They did not generally use tests but rather interviews and biographical material. In their hands the creative person came very much alive. But they often tended to rely on anecdotal material that could not be sufficiently verified or was too exceptional to serve as a basis for generalizations. The interest of these researchers was naturally caught by outstanding individuals, above all prominent authors, artists, or scientists, often deceased long ago. They did not really ask themselves if creativity could thrive under more humble, everyday circumstances.

Our own approach is not uninfluenced by these predecessors. Like the Binet followers we depend on laboratory techniques but, like the clinical researchers, we are more interested in personality than cognition and do not exclude the use of soft data. But the theoretical model guiding our view of human beings, their personality, their conception of themselves and the surrounding world is basically different.

3.2 A MODEL FOR CREATIVE FUNCTIONING

In conventional textbooks the individual is often described as a machine processing the profuse amount of information to which he is exposed. Depending upon its construction the machine may handle the information efficiently or inefficiently. This picture is turned upside-down in our model where the seemingly independent physical world around us is created from within ourselves, through ultra-short processes, mostly developing beyond our immediate awareness.

The micro-processes are rooted in our own private, unconscious experiential world. As a process continues, however, it gradually releases itself from its subjective origins. Or to express it differently, in the beginning the process is relatively free from what the American neurologist Jason Brown has called sensorial constraints, i.e., it is not strictly bound by the objective testimony of our eyes and ears. But the scope for subjective process contents is rapidly diminishing, like in a funnel turned upside-down. Finally, the process is compressed to produce a fairly objective picture of outside reality.

It is an unshakeable conviction in most normal people that the outside world is independent of the perceiver. But since it grows out of us, it must necessarily be colored by its origin, by our private world, our previous experiences, even by primitive functions that usually lay dormant in adults.

This connection between present and past can be silted up in some people while in others it remains open, even in old age. The more open it is the closer we feel to the world in which we live, the more engaged in what we are doing. At the same time, we are bound to suffer some degree of anxiety. Our existence becomes more problematic because past problems are continually being reactualized.

Figure 3.1 : One of the pictures used in the test of creative functioning

The crucial thing in this context is that, by the help of a special experimental technique, we are able to reconstruct the micro-processes by means of which our existence is continually shaped. This is not the occasion to dwell on technical details. Suffice it to say that the processes are fragmented and prolonged by means of initially brief stimulus presentations that are being gradually prolonged. Thus we make it possible to follow a process from its beginning to its end-point.

The relation to creativity is achieved by running the technique backwards. When we have adapted the subject to the so-called correct meaning of the stimulus picture we start to erode the basis for his perception by gradually diminishing the exposure times.

The fundamental problem in our creativity research is *how bound the perceiver is by the illusion of a stable outside world*, how much he takes this world for granted. One alternative for him is to preserve the illusion as long as possible because that will make him feel secure.

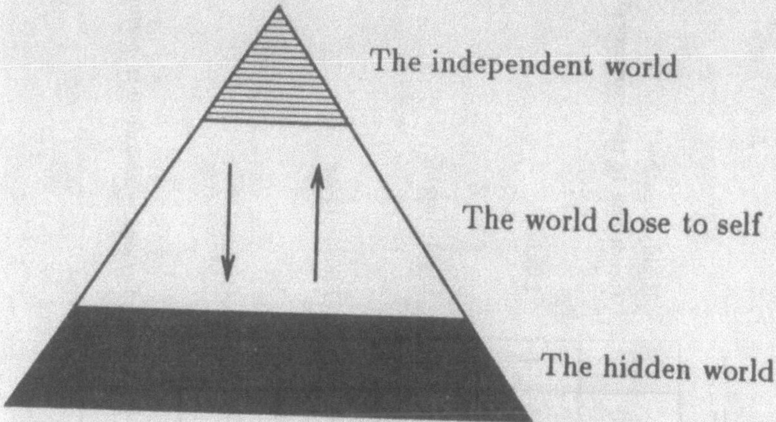

Figure 3.2 : Jason Brown's triangle

Even when presentation times are shortened to such a degree that hardly anything can be seen on the screen he will continue to refer to the established stimulus meaning: 'It's still a bottle and a bowl, has to be'. The other alternative would be to abandon the established meaning and start playing with subjective interpretations.

If we can give up the correct designation for the benefit of alternatives, less safely anchored in the sensory channels of our visual system, we may also be more inclined to renew our approach to problems presently engaging us, be they strictly personal or more public. Or to use a traffic metaphor: when reaching a dead-end street we will reverse the car and return to previous, more private functions from which we can start anew. That is, in our view, a creative way of functioning.

Before proceeding to more concrete data I would like to elucidate our thinking by means of the triangle used by Jason Brown (1988) to illustrate psychological processes. The top of the triangle depicts the independent outside world. The ascending arrow represents processes creating this world from within ourselves. As the process continues its river bed becomes more and more narrow. The bottom of the figure designates the world about which we are seldom aware; in the middle, the world which has not been released from our private self. The further down, the stronger the elements of affects, dreams, bizarre whims, i.e., experiences that are free from sensorial constraints and are often close to our personal past. A creative person is more inclined than an uncreative one to exploit the lower layers of his experiential world, not being restricted to the top.

	Highly creative (according to the experiment)	Efficient	Highly intelligent	Rich in ideas	Original	Personally absorbed in research
I	0	+	+	0	0	0
II	+	0	+	+	+	+
III	−	0	0	−	−	0

I :The efficient group
II :The creative group
III :The ordinary group

\+ Positive correlation
0 No correlation
— Negative correlation

Figure 3.3 : Picture of groups of researchers

3.3 RESULTS

I would now like to illustrate my rather abstract introduction with empirical data, mostly taken from a project recently accounted for (Smith & Carlsson, 1990).

3.3.1 Researchers

There were 18 humanists and biologists in this study, all of them relatively young but with considerable research experience. Apart from the experimental results data consisted of scaled estimates by senior professors. A personal interview was an important element in the investigation.

The diagram shows the main results of a so-called inverted factor analysis where persons, not variables are clustered. Two of the subgroups, I and II, are highly intelligent. Members of subgroup I are also notably efficient whereas members of subgroup II stand out as original and rich in ideas. The latter also receive high creativity scores in the test. Members of III, the ordinary subgroup, appear mediocre in comparison.

A few additional characteristics may help to deepen our understanding of the subgroups. The creative researchers, more often than the others, reported early childhood memories, not vague reminiscences but concrete, sensory impressions, like the heat of a sunny summer's day, the color of mother's apron when working in the kitchen, impressions often dating back as far as the age of two years. Typical was these subjects' closeness to the world of dreaming. They also were more prone to feel moderate anxiety and reported occasional nightmares.

When asked about the origin of their research problems members of the efficient group were more inclined to refer to sources outside themselves, like impressions from a journey, or just to their collection of data. Members of the creative group, on the other hand, were aware of the inner origin of their ideas. Consequently, they were more apprehensive about what might happen to the flow of ideas in the future when they became older and, as they saw it, more rigid and empty.

3.3.2 The Problem of Anxiety

In order to learn more about the role of anxiety in creative functioning we ran a separate experiment. A group of 43 people, creative as well as uncreative, was divided randomly into two subgroups. In one of them a threat was introduced in the creativity test, but outside the viewer's awareness. In this manner we planned to increase the viewer's level of anxiety. In the other group the 'subliminal' stimulation was non-threatening.

Figure 3.4 shows that the threat indeed had a positive effect. Those affected could best be described as moderately creative.

Anxiety is not always a positive factor, however. A special study of clinical subjects demonstrated that excessive anxiety can be detrimental to creative functioning. Another enemy of creativity is depressive retardation, rather self-evident when you come to think of it.

Moderate anxiety may be a necessary consequence of creative work because the process of creation, above all its inspirational phase, easily actualizes dormant conflicts and problems. At the same time anxiety is probably a promoter of creative functioning, augmenting the inner tension and raising the necessary creative temperature.

While creative people report more anxiety they seem to be more tolerant of inner discomfort, probably because they have tools to deal with contradictions, transforming them to representations at a symbolic level. The impression is sometimes that they identify with problems in order to feel more alive.

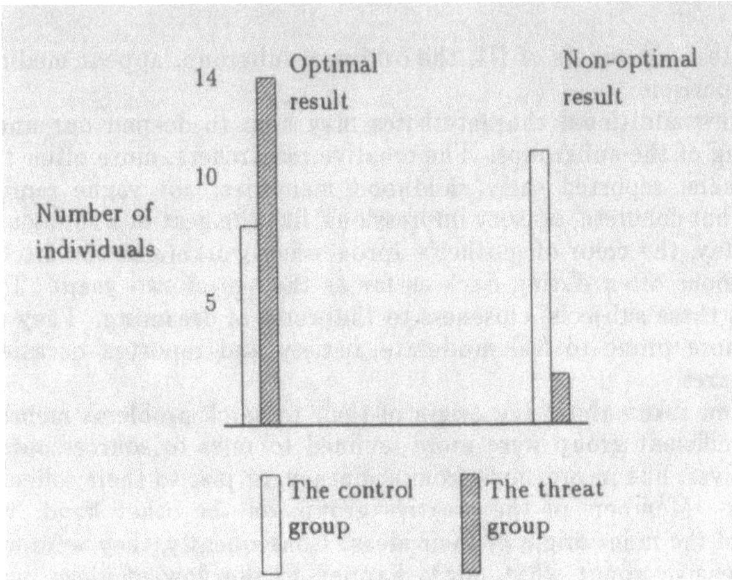

Application of a subliminal threat
in the creativity experiment
(Subjects without any trace of
creativity excluded in the figure)

Figure 3.4 : Subliminal threat and creativity

3.3.3 Creativity in Childhood and Adolescence

There is no more urgent task in the study of creativity than to describe its origins in early childhood. If creativity is defined as an interplay between the subjective and objective sides of our existence, it can hardly begin to flower until the child has developed a reasonably stable conception of a distinction between its own self and outside reality. In my homeland, given its relatively frosty human climate, the first reliable signs of creativity can be spotted at the age of five. Before our children enter regular school at 7 there even seems to be a first creative optimum.

Upon entering school, however, the children loose much of their urge to create. It returns again a few years later but disappears just before puberty, at the age of about 12 when most youngsters are extraverted, a little compulsive, and free from anxiety. Thereafter we notice a slow recovery of creativity.

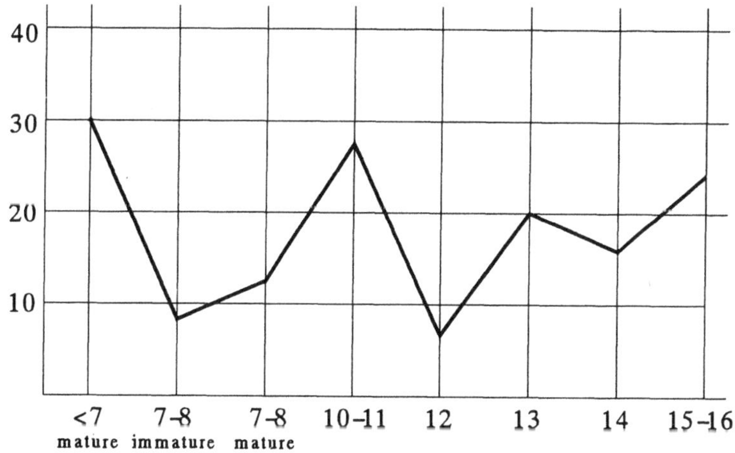

The number of creative children and adolescents increases cosiderably if moderate signs of creativity are included, even at the age of 12 years.

Figure 3.5 : Creativity in Childhood and adolescence

How can this be explained? It is difficult to blame the school system for the first decline since a recovery occurs when the children have spent still more years within the same walls. I am inclined to believe that the curve represents a natural rhythm. During the low-creative periods the child is bent on acquiring and consolidating new cognitive skills, like reading, writing, and doing arithmetics. In the high-creative periods these skills are exploited for creative purposes. Each new creative optimum represents a more advanced creative style.

It is instructive to learn that creativity can be predicted from the 5-6-years optimum to the 10-11-years one. And the best, but not the only, predictor is whether you had a fantasy playmate when you were a child.

The influence of background variables is not negligible, however. We have found, for instance, that children and adolescents from homes with a positive attitude to the intellectual and cultural facets of life are less reluctant to reinterpret the test stimulus. Therefore, they receive higher scores than those who come from culturally more impoverished families. These data are in agreement with findings among adults: individuals judged to have low self-confidence often offered uncertain

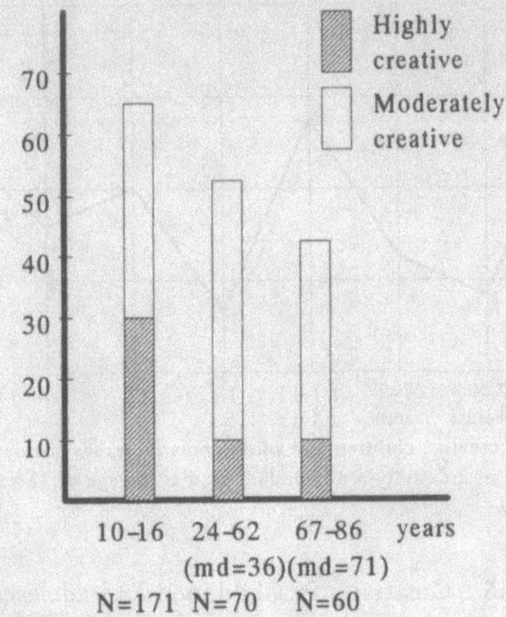

Figure 3.6 : Creativity and ageing

interpretations and were thus classed as medium-creative rather than high-creative. However, not only were the high-creatives self-confident, the uncreative ones were too.

It would be natural to ask what happens to our creativity when we go from adolescence to adulthood. We have just obtained some new data (Smith & van der Meer, 1990).

I believe that our diagram is fairly reliable. The big drop in high or accentuated creativity occurs between youth and middle age, even young middle age. After that nothing much happens up to the seventies. There is room for speculation here. To grow up and become adult is perhaps to loose the childish self-image necessary for feeling awe over life. The middle-aged seem to become less sure of themselves; they go in for cautious reinterpretations rather than bold ones. There will be reason to return to that theme in a little while.

3.3.4 Creativity and Aggression

American researchers in the field have alleged that creative people are generally more pushy and aggressive. They have not received unanimous support for this claim in other countries. As just pointed out, Americans are inclined to confuse success with creativity. And successful people may have sharper elbows than others.

We tried to test the association of creativity and aggression in our own particular way. A drawing depicting two figures, an aggressor and his victim standing facing each other, was flashed on a screen before the subject, long enough for him to describe the figures correctly. The presentations were repeated a number of times, but now with the word 'I' flashed on top of either the aggressor or the victim, the viewer being unaware of the manoeuvre.

The viewer's task was to describe the picture after each presentation. We thus asked to what degree he was ready to identify with the figure upon which the subliminal 'I' was flashed. Did he, for instance, detect the aggression more easily when 'I' was presented on the aggressor or, vice versa, when it was presented on the victim? Was he generally more positive in his description of one or the other figure, etc.? We first tested a group of 15 to 16 year-olds, then crossvalidated the findings in a group of 11 to 12 year-olds. The outcome was clear: creative youngsters were more ready to identify with the aggressor. But that does not necessarily mean that they are more aggressive. We believe that the results rather reflect a wish to be autonomous. And in the aggressor-victim dyad the former is the one who decides what is going to happen next.

3.3.5 Creativity in Artists

It may not seem particularly pertinent in this context to talk about artistic creativity. But a few results may be of general interest. First of all, artists are not necessarily creative. Some of them, as Kandinsky has said, should rather be called virtuosi in well-known genres than innovators. There was thus a considerable dispersion with respect to creativity in the group of professional artists serving as our subjects. It could also be mentioned that, when tested with a method reminiscent of the one just described, creative artists preferred an open, loosely structured background to a closed, tightly structured one; that they were more inclined to see their childhood years in both positive and negative colors than to describe them in an entirely nostalgic or misanthropic perspective. All this is of interest in view of our thesis that creative functioning grows out of complications and flourishes in an atmosphere of moderate uncertainty.

3.3.6 Creativity and Self Image

How do creative people regard themselves? In order to illuminate this problem we flashed an ambiguous face on a screen in front of the subject, long enough for the face to be recognized as a face but still briefly enough for the subject to project himself onto his description of it. Short word combinations like 'I BAD' and 'I GOOD' were exposed 'subliminally' on the face in order to manipulate the viewer's identification with it.

Creative people were more inclined to vary their characterizations of the face. These variations concerned such things as mood and sex. But most interesting were the changes in age. To creative people the face would sometimes represent an adult, sometimes a child or an adolescent. If, as we have good reasons to assume, the face descriptions reflect the viewer's own self-image, creative people thus have a double perspective, the curiosity and irreverence of a child combined with the critical distancing of an adult. This is also true of elderly people, at least to some extent. It is of particular importance that a troublesome subliminal message like 'I BAD' is more efficient in this respect than the soothing 'I GOOD'.

3.3.7 Creativity and Aging

Creativity seems to be decisive for our attitude to aging. We asked 70-year-old subjects to read the health journals of two elderly ladies, one who had remained reasonably healthy all her life in spite of disturbing symptoms in her youth and another who had a long history of relapses into periods of illness. A number of explanations for their respective conditions were offered and the subjects were asked to pick out the four most likely ones.

Creative people preferred alternatives where the subject had actively tried to help herself, or failed to do so, and were more inclined to acknowledge psychological factors than deterministic-biological ones. They were not willing to delegate all responsibility for their health to doctors and their medicines.

On the whole, the creative elderly were more satisfied with their present condition, regarding the remaining years, however few, as a resource to be used in a productive way. They were less ready to deny the possibility of a life after death.

Uncreative people were more uniformly stable. Their self-confidence often seemed as unshakable as that of highly creative subjects. But it was a confidence of a different kind, based on the absence of psychic complications. They more often talked about how to *pass* the time than about how to *use* it.

Table 3.1 : Summary of findings

Characteristics of non-creative individuals
Strong anxiety or no anxiety
Depressive retardation
Compulsiveness, i.e., preference for rigid rules and order, dread of feelings
Intolerance of inner discomfort
Unability to, or fear of internalizing problems and conflicts
Fear of solitude
Conventionalism
Inability to feel awe
Characteristics of creative individuals
Moderate anxiety
Flexibility, an urge to reconstrue reality
Openness towards the inner landscape
Tolerance of inner discomfort
Unwillingness to understand the good of rigid rules and order
Internalization of problems and conflict as the starting-point for creative endeavours
Autonomy and need for solitude
Ability to feel awe

3.4 SUMMARY, CONCLUSIONS, AND SPECULATIONS

Table 3.1 is an attempt to summarize the basic characteristics of creative and uncreative individuals, ignoring the sizeable group in-between these extremes. A creative individual builds his existence from within himself, including not only the well-adapted aspects of his inner life, but deviating, childish, ridiculous aspects as well. Even after he has grown up these aspects remain accessible to him. To be ashamed of them, to ban them from his conscious life is to shut up one of the main sources of creative functioning.

Maybe creative people are less complete, less established and grown-up than uncreative ones. It is obvious to all who have studied creativity that creative people are more filled with contradictions and conflict, often more tormented, than uncreative ones. At the same time they seem more tolerant of discomfort. They are equipped to handle conflicts in a constructive manner. We should not deceive ourselves that creative people are necessarily more happy, or more entertaining in a superficial sense, some kind of cocktail party heroes. They can be unsociable, sometimes unwilling to adapt. Neither need they be particularly talented. But I dare say that they lead a more meaningful life because they can still feel wonder at their existence.

Since creativity came into fashion smart people have tried to sell special training courses. I have always been skeptical about such undertakings. Judging from our own studies, however, it would not be impossible to do something for the middle group of overcautious, moderately creative people, to make them more willing to invest interest in ideas that may at first seem abstruse and unrealistic but, upon closer scrutiny, could perhaps open up new vistas.

Everybody wants to speculate about the role of environmental factors for the development of creativity; Elliott Jaques (1990) has written extensively on the importance of the work situation. In most speculations cultural pluralism has been considered as a positive factor for promoting creativity, in contrast to cultural homogeneity and the dead weight of tradition. It seems more or less self-evident that a child forced into rigid rules and fostered to dread breaking them, i.e., a child exposed to an authoritarian upbringing, could not possibly grow up to become a creative person. This does not mean that I see total absence of rules as a necessarily positive factor. Creating is a process of dialectics between private whims and established ways of thinking. We need a wall for our ball-playing, a wall named knowledge. The important thing is not to respect that knowledge just because it is established, only if it is good and useful.

REFERENCES

Brown, J.W., 1988, *The Life of the Mind, Selected Papers,* Prentice Hall, Englewood Cliffs, N.J.

Jaques, E., 1990, *Creativity and Work, Emotions and Behavior,* Monographs 9, International Universities Press, Madison, CT.

Smith, G. J. W., and I. M. Carlsson, 1990, *The Creative Process,* Psychological Issues 57. International Universities Press, Madison, CT.

Smith, G. J. W., and G. van der Meer, 1990, "Creativity and Old Age", *Creativity Research Journal 2,* 3:249-64.

CHAPTER 4

Synergetics as a Theory of Creativity and Its Planning

Hermann Haken

4.1 NATURE AS THE GREAT TEACHER

Quite often the word 'creative' refers to a single person who produces outstanding kinds of work, such as music, paintings, poetry, technical innovations, or scientific ideas. How a creative person makes his or her achievements is one of the great mysteries of the human mind. In the context of this Chapter, the meaning of the word creativity encompasses a much wider scope. It refers to the creativity of a whole community, not only a city, but even a large important area. While it is practically impossible to plan the creativity of a single person, planning that of a whole community may be feasible. This is because a large group of people may produce ideas or products by their cooperation. This then leads one to the question where we can find guidelines or paradigms for organizing the creativity of large groups. Such guidelines may possibly be found if we look at nature.

Nature herself impresses us with her enormous creativity which has produced an overwhelming variety of species with remarkable features. Let us therefore conceptualize a city or an area as a giant organism. Indeed, even at a rather superficial level we may establish a number of analogies between the action of an area and an organism.

Any organism is an open system with its uptake of energy and matter. The same is true for any city or area. Matter is imported and exported in the form of raw materials and products, such as ships, cars and computers; energy is also imported and exported, in the form of oil, coal, uranium, electric currents, etc. What is becoming increasingly important is the importation and especially the exportation

Table 4.1 : Analogies between the action of an area and an organism

organism	city
blood circulation	traffic
food uptake	input of goods and energy
digestion	transformation of energy and work
nervous system	information networks
brain	administration, technology centres, etc.
sensory organs	information up take
locomotion, work done	products
waste disposal	waste disposal

of information which is based on basic research and applied science involving inventions, scientific discoveries, patents and publications. Applied science and technology are playing an increasingly profound role. There is an increasing intellectual demand for high-level TV programs, computer software, etc. As was stated some years ago by ÅE. Andersson (1993), quite clearly the trend is towards an increasingly information-intense form of technology in which the percentage of matter and energy becomes less and less important as compared to information. In the long run, this will have important impacts on the organization of cities and areas. In earlier times the processing of coal and steel, for instance, required short distances between the sources of the raw materials and their processing sites in order to keep costs low. On the other hand, when the main emphasis lies upon the development of software, for example, the processing can be done in private homes which may be quite distant from administrative centers, etc.

Quite clearly, creativity is intimately linked to increasingly highly valued information - to its production, processing and transference. In order to study how creativity can be planned and organized, we can examine nature's basic principles for achieving her creativity and efficiency. Below, I list some of the features of these principles.

(1) Biological systems possess an enormous degree of coherence or coordination. Practically all activities, say of a human or an animal, require a high degree of coordination of an extremely great number of parts. In locomotion, breathing, heart beat, or blood circulation a very great number of cells must cooperate coherently. Speech requires the cooperation of billions of

nerve-cells, muscle-cells, and tissue-cells. At a still higher level thinking requires an incredible degree of cooperation between neurones. Nature has managed to transform microscopic order or even microscopic chaos to macroscopic order.

(2) Nature has managed to maintain a delicate balance between reliability (stability) and adaptability.

(3) Nature has developed the means to explore new possible forms of life, settling down into ecological niches, and developing new kinds of species.

(4) The processes in nature are not imposed on the individual organisms from the outside; they develop by means of self-organization.

4.2 LASERS AND FLUIDS AS PARADIGMS IN SYNERGETICS

As has been revealed by synergetics, a number of the properties listed above can also be found in systems of the inanimate world, which allows us to study the basic principles of self-organization more explicitly. These principles have been used to model and understand numerous processes in biology, e.g. the coordination of muscle activities leading to locomotion and movements, pattern recognition by the brain, the evolution of species, population dynamics, the formation of public opinion amongst humans, etc. It may be worthwhile to illustrate these principles by means of a simple example taken from physics.

Such an example is provided by the light source laser. A gas laser consists of a glass tube filled with a gas comprised of individual atoms. When an electric current is sent through the glass tube, the current may excite the individual atoms, where upon each atom acts as a small radio antenna emitting a light wave, which is just a radio wave with a very short wave length. In a common lamp these individual waves are emitted in a chaotic fashion. It is as if we throw a handful of pebbles into water and generate a wildly excited water surface. From a superficial point of view, the laser is only distinguished from a lamp by means of two features. There are two mirrors mounted at the end faces of the glass tube, which serve to reflect the emitted light quite often so that it can interact with the atoms for a longer period of time. Furthermore, the electric current is enhanced as compared to that injected into a common lamp. But now an entirely new phenomenon occurs: The atoms no longer act independently of each other; rather, they emit a highly ordered wave, the so-called coherent laser wave.

When such an ordered wave is emitted, the efficiency - i.e., the ratio between the power of the emitted laser light and the power of the input current - is greatly enhanced.

The atoms have found the internal mechanism of coherently emitting light by means of self-organization; i.e. there are no external means imposed on the system in order to make it act coherently. In the parlance of synergetics, the emergent laser light wave acts as an order parameter. It prescribes the action of the individual atoms by means of the slaving principle of synergetics. According to this principle, slowly varying quantities, such as the laser light wave, may enslave rapidly adapting subsystems, such as the atoms. While, on the one hand, the order parameter determines the behavior of the subsystems (of the atoms), the atoms, on the other hand, support the laser light. In this way we find circular causality between the order parameter and the subsystems.

Another well-known example of self-organization is provided by fluid dynamics. When a fluid is heated from below, it may spontaneously form specific patterns, such as honeycomb structures. In each honeycomb cell the liquid rises, cools and then sinks down the cell's borders.

4.3 BASIC CONCEPTS AND PRINCIPLES OF SYNERGETICS

These principles can be summarized as follows: We consider complex systems to be composed of many individual parts interacting with each other. The systems are open, i.e. their action is maintained by a continuous input of energy or matter, or of both. In spite of the fact that the individual elements may be quite different, the macroscopic behavior of such systems shows striking analogies. When certain control parameters such as energy input or social climate are changed, the system may undergo qualitative changes. The macroscopic structure may consist of specific spatial or temporal arrangements, or special functions. The efficiency, in particular, can be improved quite considerably. When such macroscopic structures occur, a high degree of coordination is reached among the actions of the individuals. This is achieved by self-organization, in which the individuals generate one or several so-called order parameters which in turn influence ('enslave') the action of the individuals. Such order parameters may be the information flux generated in the system, or energy fluxes and so on. Note that this information transfer is in general not achieved by a central agency. Rather, the order parameters represent the collective state of the system. Thus, the well-known information bottle-neck observed in some highly centralized countries is avoided.

When the control parameters are changed further, the system can run through a hierarchy of different highly ordered and coordinated states. Simple examples are provided by physical systems, such as the laser and fluids. When more and more energy is pumped into a laser, it may run through the following different states: that of the microscopically chaotic light of a lamp, the coherent laser wave, regular ultra-short light flashes, and so-called deterministic chaotic laser light. Similarly, in some fluids the following may occur: first a fluid is at rest, then motion begins in the form of rolls, then the rolls start to oscillate, and then they show a more and more complicated oscillation. Under self-organization a system may acquire different states. That which is selected depends on adequate initial states and on additionally imposed constraints. This can easily be demonstrated by means of a fluid heated from below. When the fluid is in a circular vessel, a motion in the form of rolls (or stripes) starts. The direction of the rolls may be arbitrary and depends on the initially given order of the system. When the circular vessel is replaced by a quadratic one, only two roll directions are still possible; that which is chosen depends on the initially given state. Finally, when the vessel is no longer square, but rectangular, only one roll pattern may emerge. All these phenomena can be understood by a rigorous mathematical theory that was developed within the framework of synergetics. This theory can be sketched as follows: The microscopic state of the system is described by the so-called state vector. In a physical system, such as the laser, it may be the electric field of the laser light, the polarization of the atoms, and so on. In sociology, it may involve the number of persons with a specific job, or the number of factories with a specific product, etc. The evolution of the state vector over the course of time is described by a set of nonlinear differential equations, where the time evolution depends on the present state of the system. The mathematical formalisms then allow us to determine those control parameter values in which the system changes its behavior qualitatively. It further allows us to determine the order parameters and to ascertain those types in which the subsystems are enslaved by the order parameters. Using the slaving principle of synergetics it becomes possible to reduce the enormous complexity of a system with its very many variables to a few variables, namely the order parameters.

In contrast to physical or chemical systems, in which the future of the system is determined by its present state, in human society another important feature is present, namely the role of expectations about the future. The actions of people are not only influenced by the present status of the world, but by their hopes and fears or economic expectations. Thus, it is not only material circumstances which determine the build-up of specific order parameters, but also the hopes generated among people. It is important to note, however, that the concept of order parameters still remains valid as well as the slaving principle. One may even state that expectations about the future play the role of order parameters. For instance, when there is an optimistic view about the future development of the economy in a society, it may well be expected that the economy will develop strongly, because the consumers and investors behave in an according manner (they are 'enslaved' by the optimistic view).

4.4 SOME GENERAL CONCLUSIONS

Let us now try to draw some general conclusions from the concepts of synergetics with respect to the planning of creativity in an area. Let us start with a model; namely, the establishment of a laboratory for research and development. Of course, we want to have a highly creative laboratory. We need highly innovative co-workers who are highly motivated. Furthermore, we expect them to be cooperative. How can we attract such people? First of all there must be a selection process, e.g. by entrance examinations or supporting letters from their professors, etc. We must make their working conditions as favorable as possible; e.g. by adequate rooms, adequate equipment, and so on. But their private lives must also be as favorable as possible. There must be adequate schools, shopping centers, recreation centers... Depending on information technology and various other requirements, the work may be centralized or decentralized to private areas.

But above all, we must set up adequate order parameters. These are corporate identity, a favorable working climate - which must be built up slowly by means of reliable personnel - willingness to exchange ideas, a common goal for the company. It must be observed that the order parameters are durable quantities; they may be of an immaterial character, such as the social climate within a company, or they may be of a material character, such as buildings, communication channels, etc. Quite clearly, the kind of architecture also acts as an order parameter. The kind of order parameters chosen will determine the character of self-organization and there are limiting cases. For example, one order parameter may allow complete liberalism in the creativity of the individuals; another order parameter may mean that the work of the individuals is highly regulated. This may depend, indeed, on the kind of task to be fulfilled. It is not too difficult to scale this model up to the planning of a whole area.

Lack of space does not allow me to present a fully developed theory of the planning of a city or of an area. I wish to mention that synergetics allows us to construct dynamical systems which produce any desired pattern once appropriate connections between the parts of the system have been established and once adequate initial conditions are provided. These construction principles could be realized on the synergetic computer and are again based on the concepts of order parameters.

4.5 CONCLUDING REMARKS

Let us discuss another aspect, namely the problem of adaptability versus stability. Generally we assume that it is important for the performance of a system for it to have a high degree of stability. However, in a rapidly changing world it is more important for a system to be adaptable. That means it is wise to keep it close to a point of instability; from there it can be driven rather quickly to new points of action. There is increasing evidence that the human brain acts along these lines.

As one may show by means of the results of synergetics, it is more advantageous to produce generally favourable conditions than to produce over-detailed plans. There must be enough space for initiative by individual people, who then cooperate on a large scale and are driven by a common goal - namely, to contribute their best towards the prosperity of their own area. This common goal will then increase the quality of life of that area significantly. On the other hand, it must be recognized that synergetics teaches us that occasionally self-organization will drive a system into a local state at a certain level of efficiency, but from which the system cannot escape into a state of higher efficiency. It is here where governmental decisions or agreements between companies must be made in order to bring the system in a coherent fashion to that state of higher efficiency. But here again too much control must be avoided. Rather, the art is to find as few and as efficient control parameters as possible to drive the system from one such a state to a better state.

So let me conclude by making the following remark: The creativity of an area will principally be based on two factors: a high degree of initiative by individuals and the production of coherent patterns of activity.

REFERENCES

Andersson, Å.E., *Economic Structure of the 21st Century,* in this Volume.

Haken, H., 1983, *Synergetics, An Introduction, 3rd ed.,* Springer-Verlag, Berlin, New York.

Haken, H., 1991, *Synergetic Computers and Cognition,* Springer-Verlag, Berlin, New York.

CHAPTER 5

High Technology Worker Mobility

Christer Anderstig and Björn Hårsman

5.1 THE GROWING KNOWLEDGE ORIENTATION

The most important institution and the driving force of Francis Bacon's utopian society Bensalem was a scientific foundation called Salomo's house. Hopefully, science will never become as dominant as Bacon prophesied. However, the ongoing structural transformation of the world economy certainly indicates a strongly increasing knowledge orientation.

The transition from an industrial society to a service and knowledge society seems to be a characteristic trait of most industrialized countries. Andersson (1986) offers a broad and historical perspective on this transition and introduces the concept 'logistical revolution'. This concept is meant to reflect the combined effects of the computing and communications revolution and the successive growth of the knowledge base and the transportation system. The impact of this increased knowledge orientation is assumed to rival that of the replacement of muscle power by machines during the industrial revolution of the 18th century.

The regional development which has occurred in Sweden during the last 30 years seems to be in line with Andersson's hypothesis about the importance of a region's accessibility and its knowledge base. Anderstig and Hårsman (1986a) show that knowledge intensive activities are becoming more and more important in the Swedish economy. As illustrated by Figure 5.1 these activities tend to grow faster in large regions; i.e., in regions that are usually characterized by good accessibility and by a strong knowledge base in the form of universities and colleges. In Figure 5.1 the increased knowledge orientation of certain regions has been measured by means of two occupational groups.

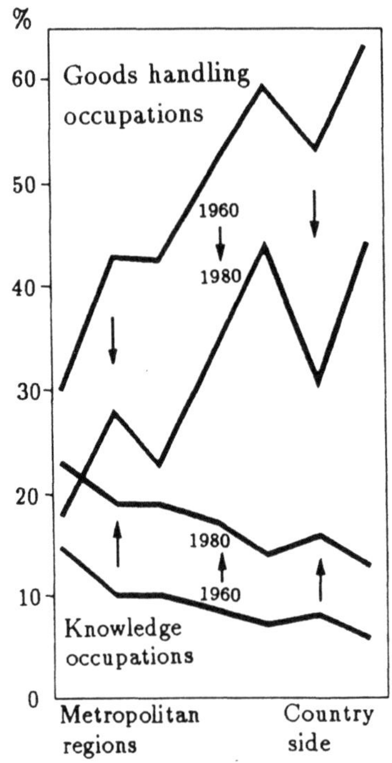

Figure 5.1 : Percentage of employment absorbed by knowledge occupations and goods handling occupations in seven different Swedish regions, 1960-1980

Knowledge occupations include technical and scientific work (physicists, chemists, engineers, physicians etc) and other knowledge based work (social scientists, teachers, journalists, lawyers, judges, artists etc.). Goods handling occupations include farmers, fishermen, craftsmen and all kinds of workmen involved in manufacturing, construction etc. The seven regions in Figure 5.1 have been defined according to intra-regional accessibility.

Figure 5.1 demonstrates that the knowledge orientation has increased in all regions and that the metropolitan regions are about 10-15 years ahead of the other regions in this respect. In Stockholm, the largest metropolitan region, almost 25 percent of the total employment is currently absorbed by knowledge-related occupations and about 15 percent is absorbed by goods handling occupations.

The Stockholm Region is by no means unique in respect to knowledge orientation. The following Table, from Andersson et al. (1988), compares the structural change which has occurred in the Stockholm and Tokyo Regions between 1970 and 1980. Knowledge and goods

Table 5.1 : Occupation structure in the Stockholm and Tokyo Regions, 1970-1980 (Employment change per year in percent)

	Knowledge	Information	Services	Goods
Stockholm	+2.8	+0.9	+2.3	-2.0
Tokyo	+2.2	+0.5	+1.4	-1.6

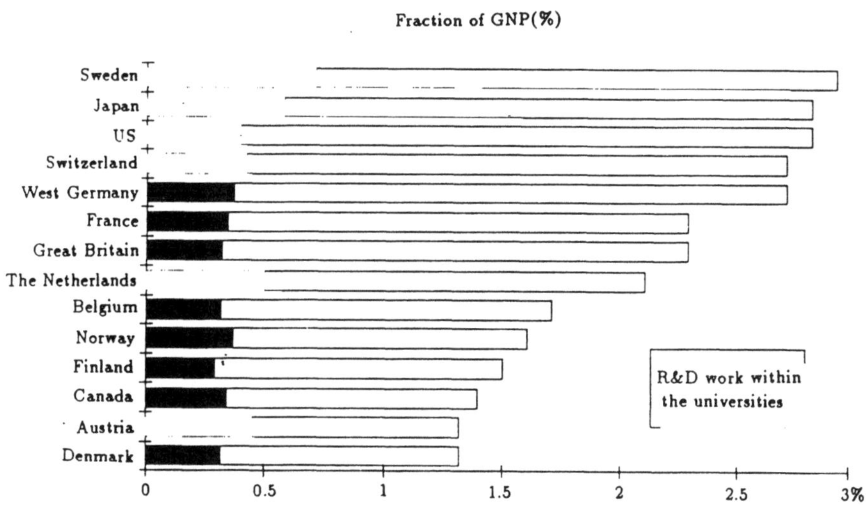

Figure 5.2 : R&D expenditures as a percentage of GNP in certain OECD countries, 1985

handling occupations are defined as above. Information occupations include managers, accountants, cashiers, secretaries, typists etc and service occupations include nurses, clergymen, tradesmen, bus drivers, policemen, cooks, barbers etc. Table 5.1 demonstrates that the patterns of change are similar in Tokyo and Stockholm. Figure 5.2 illustrates some similarities between Sweden and Japan in regard to

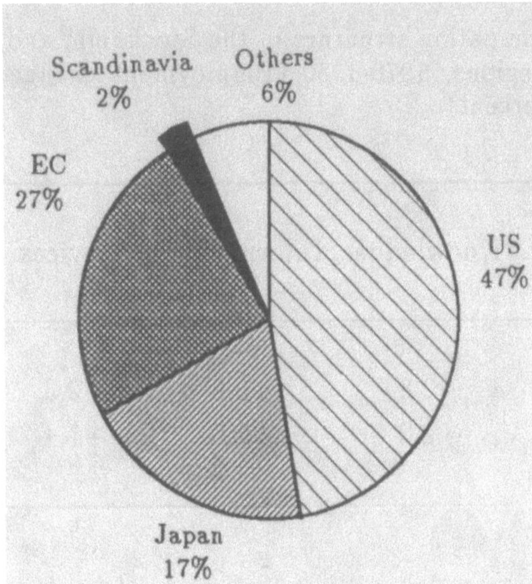

Figure 5.3 : Allocation of R&D expenditure within the OECD

another aspect of knowledge orientation: it shows R & D expenditure as a percentage of GNP in certain OECD-countries. Sweden occupies a leading position in this respect and is closely followed by Japan and the US. Figure 5.2 also shows that about the same percentage of the R & D work is performed within the university system in Japan as in Sweden. In fact, Japan is the only large country with more than 20 percent of its R & D work performed within the university system.

The total R & D expenditure in the OECD countries was about 230 billion dollars in 1985. Sweden's share was 3 billion dollars; i.e., 1.3 percent. Sweden's small size is further illustrated in Figure 5.3, which shows the allocation of R & D expenditure among the OECD countries in 1985. The US accounts for the largest share and Japan for the next largest share of R & D expenditure. West Germany (9%) ranks third and France (6%) and Great Britain (6%) equal fourth. Together these countries account for 85 percent of the R & D expenditure within the OECD.

5.2 THE IMPORTANCE OF INFRASTRUCTURE

The growing knowledge intensity implies a growing importance of skilled labour and of R & D resources. This development is partly a result of the shift from manufacturing to services. But the knowledge orientation of the labour force has also increased in manufacturing. By way of example, the percentage of manufacturing employees in knowledge oriented occupations doubled from 13 percent to 26 percent in the Stockholm Region between 1960 and 1985.

The increasing knowledge intensity in manufacturing is basically related to the computer and communications revolution. Computerization and telecommunications have increasingly made it possible to coordinate production and the transportation of commodities over long distances. They have also led to a shift from a hierarchical organization of production activities to a more flexible network organization. Within firms various activities can be allocated to different establishments within the same region or in other regions. An additional reason for this is that technical progress has reduced the optimal scale for many types of economic activities.

As a consequence of this development, infrastructure - in the form of roads, railroads, airports and universities - has become more important. Another consequence is the changing pattern of business locations. Traditionally, manufacturing firms have chosen locations adjacent to 1) the source of their raw materials, 2) their markets or 3) nodes in the transportation network.

In a knowledge oriented society locational choice is better explained by the theory of product cycles. According to this theory the centre of production and employment is shifting inter-regionally or internationally as products develop from high R & D intensity to maturity. It can be envisaged that metropolitan areas have a locational competitive advantage in the first phase of the cycle, dominated by research, knowledge oriented activities and technical and commercial development. Against this background one would expect knowledge oriented firms to locate in regions with a large supply of skilled labour and R & D resources, and a high level of inter-regional and intra-regional accessibility.

The importance of a large supply of skilled labour and R & D resources is confirmed by the results from surveys among high-tech US firms (see Premus, 1982). According to this study the supply of skilled labour is the most important factor of location both at the inter-regional and the intra-regional level. To be located adjacent to universities is also ranked very high. Quigley (1990) has analyzed the combined effects of the changing production technology and trends in housing demand caused by the increased female labour supply, rising household incomes and reduced transportation costs. He concludes that 'firms seeking to attract highly skilled or highly qualified workers will be much more likely to choose the metropolitan area that offers a higher quality housing stock, lower overall housing prices, or a higher level of housing amenity'.

The importance of infrastructure is also evident from empirical

Table 5.2 : Results of five different model applications

	Model				
Explanatory factors	1	2	3	4	5
Population	+	+	+	+	0
Travel time		0	0	0	–
Airport			+	+	0
College or university				+	+
College and airport					++
University and airport					+++
R^2	0.30	0.30	0.47	0.55	0.60

studies we have performed using Swedish cross-sectional data; (Anderstig and Hårsman, 1986b). The following table reports results obtained when analyzing differences in the knowledge orientation of 70 Swedish labour market regions. Table 5.2 presents results from five different model applications. The dependent variable is in each case the percentage of employees with knowledge oriented occupations (according to the definition indicated above). The explanatory factors are population, travel time measured as an index of inter-regional accessibility, and the availability of an airport, university or college. The + sign means a positive correlation with the corresponding explanatory factor and that the relationship is significant at the 5 percent level, ++ at the 1 percent level and +++ at the 0.1 percent level. The 0 sign means not significant and the - sign a significant negative correlation (at the 5 percent level). The same kinds of results were obtained for 1970 and 1980 and also when knowledge orientation was measured as the percentage of employees with a university degree.

Model 1 shows that size alone has a positive and significant influence on the knowledge orientation in a region. However, only 30 percent of the variation in knowledge orientation is 'explained' by this factor. The determination coefficient increases when additional explanatory factors are successively introduced. In the most complete model, i.e., model 5, all coefficients have the expected sign and the determination coefficient has increased to 60 percent. According to this model, size per se does not influence the degree of knowledge orientation. It rather seems to be the case that investments in infrastructure, which are correlated with population, explain the variation in the knowledge orientation of the regions.

The importance of infrastructure for the economic growth of a region is also demonstrated in a paper by Andersson, Anderstig and Hårsman (1990). Their study is concerned with an analysis of the relationship between infrastructure and productivity. It is based on a production function approach, permitting variable returns to scale with respect to the quantity and quality of labour. The infrastructural capacity of a region is represented by a geometric aggregate of air, road, rail, building capital and R & D capacities. A non-linear econometric procedure is used to estimate the influence of infrastructure upon the gross productivity of regions.

The econometric study shows that there are clear relations between the supply of infrastructure and the productivity of the regions as macroeconomic entities. The most important infrastructural factors are related to research and development capacity and to transportation and communication possibilities. Total production is primarily influenced by road, airport and R & D capacity.

In addition, the study shows that while railroads had a significant effect upon regional productivity in 1970, this effect does not appear in the results for 1980. Fast and flexible transportation modes are obviously of growing importance. As in the study by Anderstig and Hårsman (1986b), airline transportation and R & D capacity interact synergetically. In fact this synergistic factor turns out to be highly significant and important for regional productivity.

The importance of universities is confirmed by Jaffe (1989). He uses state-level time-series data to relate university research and corporate R & D to corporate patents. One of the conclusions is that 'a state that improves its university research system will increase local innovation both by attracting industrial R & D and augmenting its productivity'.

5.3 High-Technology Worker Mobility

As has been pointed out above, traditional location factors such as accessibility to markets and raw materials are of minor relevance for the location decisions of high-technology firms and other firms within the expansive knowledge-oriented sector of the economy.

In the traditional industrial society workers had neither the time nor the money to live far from their work-places. This makes it reasonable to encapsulate the former location pattern in the expression 'workers followed firms'. Today it is often stated that this expression should be the reverse, that is 'firms follow workers'. Although not generally valid, 'firms follow workers' certainly reflects important tendencies with respect to the hi-tech sectors in modern societies. In this context Klaassen's (1987) discussion is interesting, concisely stating: 'Modern economic activities will concentrate in areas where the living is good'.

What Klaassen has in mind are the visible patterns resulting from the residential preferences of highly-skilled manpower in Europe. 'A number of favourite residential areas in Europe are concentrated in the Alpine region, consisting of southern Bavaria, Austria, Switzerland, Savoy, Dauphine, south Baden-Wurttemberg, the Cote d'Azur, and the lake region in northern Italy. The Alpine region is a vast area with at least seven international airports and as many sizeable towns. The area has also much to offer in the way of culture. It accomodates the Salzburger Festspiele, and La Scala in Milano as well as universities and scientific institutions of world fame. Indeed our statement about the Alpine area was clearly a macro one, emphasising the importance of the living environment in a general sense. The micro preferences still have to be looked at closely. In every country there are parts where living is pleasant, and these parts will play, within their national borders, the role the Alpine region plays in the European context'.

Some empirical work has been completed incorporating a close examination of these micro preferences. In the U.S., Herzog, Schlottman and Johnson (1986) and Herzog and Schlottman (1989) have analyzed the migration decisions of high-technology workers, as well as the impact of these decisions in relation to the locational determinants of the high-technology industry. Some of their findings imply that high-technology workers are more mobile than other workers, and that high-technology migrants are distinguished from other migrants by their preference for large metropolitan areas.

To some extent Herzog and Schlottman have inspired the present Chapter since there is a mutual interest in comparing the results from corresponding analyses in Sweden and the U.S. Our approach has, however, been broader as analyses have also been made with respect to the migration decisions of workers other than high-technology workers; that is, the following occupational groups, also mentioned above:

Technical scientific jobs including physicists, chemists, engineers, architects, physicians, biologists, etc.;

Other knowledge-handling jobs including social scientists, teachers, programmers, journalists, lawyers, judges, artists etc.;

Administration and information jobs including managers, accountants, cash-iers, secretaries, typists etc.;

Personal services including nurses, clergymen, tradesmen, bus drivers, policemen, cooks, barbers etc.;

Goods-handling jobs including farmers, craftsmen, industrial workers, construction workers etc.

We include a statistical summary of the migration rates of all occupational groups at the start of the section on the empirical results of our study, although our main interest is with respect to high-technology workers. High-technology workers are mainly found in the first category above - Technical scientific jobs - although some are also incorporated in the second category: Other knowledge-handling jobs. In the definition of hi-tech workers, four sub-groups have been classified: 1) Computer specialists, 2) Engineers, engineering and science technicians, 3) Economists and statisticians, and 4) Life and Physical scientists. These groups, as far as possible, correspond to the groups defined by Herzog and Schlottman.

Our analyses are based on microdata on households for the periods 1975-1980 and 1980-1985 and data were obtained from one percent samples of the 1980 and 1985 Censuses, with complementary data from 1975 and 1980, respectively. As for the specific analysis of high-technology workers, data were obtained from ten percent samples of the 1980 and 1985 Censuses - about 21,000 households. Only results based on the 1985 data will be presented here. This microdata was processed to extract those householders aged between 24 and 60 who were members of the labour force in 1985. The sample was drawn from the total population, stratified into sub-populations according to outmigration during the period 1980-1985 (w.r.t. labour market region, 'A-region'), occupational group, and permanent address (Sweden subdivided into eight county-blocks).

The occupational group is only one of the determinants of migration. In a recent survey by Greenwood (1985) a number of critical life-cycle considerations are presented, such as marriage, divorce, completion of schooling, entry into the labour force, as well as personal circumstances such as employment status, education and sex. Our aim has been to take these determinants into consideration as far as possible. Additional personal characteristics include the employment status and occupation of the spouse, experiences of long-distance commuting and prior geographic mobility.

Table 5.3 : The migration rate for different populations and occupational groups in percent

	Occupational group					
Population	Techn. scient.	Other knowl.	Adm. info	Pers. serv	Goods-handl.	Total
Period 1975-80						
Entrants	52	39	20	22	17	26
All	12	17	9	10	6	9
Period 1980-85						
Entrants	41	33	18	21	16	24
All	10	13	7	8	5	8

In order to model the decision to remain within the A-region or to relocate over the ensuing five-year period, personal characteristics have been complemented by a set of regional characteristics. When selecting these regional characteristics, we have only in some respects used the same characteristics as in the U.S. studies, partly because some of the U.S. characteristics are of minor interest in the Swedish context due to low inter-regional variation, and partly because of a lack of data. However, some variables are more or less the same, such as those aimed to reflect general economic conditions (total employment growth and per-capita income), city scale (population), the employment share of knowledge-oriented occupations, home prices, and accessibility to transportation. As for educational quality and accessibility to cultural amenities, we have so far only applied dummy-variables for the seven cities with universities.

5.4 RESULTS

Table 5.3 provides a summary of two general data sets. For the period 1975-80 (1980-85) the sample consists of 19,018 (19,895) householders. The number of households who relocated during the period, i.e. migrated, is 1699 (1594). This means that the migration rate per five-year period for the A-regions is 9 percent and 8 percent respectively.

The occupational groups are classified according to occupation at the end of the period. Those who did not belong to the labour force at the beginning of the period are handled separately. Many of these were students who entered the labour force during this period. We call them entrants. The reason for this division is the completely different decision-making process these household categories face. Those entering the labour market have a migration rate which is about three times the average rate, and for those who entered a technical scientific job the migration rate is about twice the average. We can see that the result is quite stable over time. The structural relationships are maintained with only small variations in the parameters.

Let us now turn to the specific analyses of high-tech worker mobility. These have been performed both with respect to all the 21,190 high-tech workers 'at-risk' to 1980-85 migration, and to a sub-group of 18,591 high-tech workers who were also in the labour market in 1980, thus excluding 2599 entrants to the labour market during the period. For the total group the migration rate was 11.4 percent, while it was 8.0 percent when the entrants were excluded.

The results refer to four models. First, we have pursued the idea of separating the entrants to the labour market from the total group. Next, to make some test for the power and/or quality of the regional variables, we have introduced a version of the model where these characteristics have been dropped. Binary logit estimates of the determinants of migration are provided in Table 5.4. An asterisk (*) indicates that the coefficient estimate is significant at the 5 percent level.

First it can be noted that all coefficients related to personal characteristics are significant. The results show that among high-tech workers mobility - that is, the probability of moving to another labour market region - reduces with age and is lower among female workers. In addition, mobility rates are lower for married workers, those with a working spouse and school-age children. On the other hand, high-tech worker mobility is augmented by non-employment (e.g., unemployment), by prior migration and by commuting at the outset of the migration interval. In cases where the spouse is employed, it is interesting to note that if the spouse has a knowledge-oriented occupation, the probability to migrate is somewhat higher, compared to other occupations.

These results are similar to the results in the corresponding analysis by Herzog and Schlottman, and in several respects these determinants are also as important for other workers as for high-tech workers. The coefficients for the three occupational dummy variables indicate that the likelihood of migration during the period 1980-1985 is higher for Life and Physical scientists than for Engineers and Economists and Statisticians (the omitted occupation). These results are not really

Table 5.4 : Determinants of 1980-1985 migration of high-technology workers: binary logit estimates

Variable	All workers		No entrants	
	Reg char	No reg char	Reg char	No reg char
Constant	0.711	-0.1575	-0.1550	-0.3324
Personal characteristics				
Female	-0.3134*	-0.4472*	-0.4142*	-0.5357*
Age	-0.0697*	-0.0693*	-0.0695*	-0.0696*
Not employed	0.2581*	-	0.0848	-
Prior migrant	0.8661*	0.8048*	0.9501*	0.8690*
Married	-0.3556*	-0.2774*	-0.2839*	-0.1788
Employed spouse	-0.3775*	-0.3657*	-0.4198*	-0.4356*
Spouse with knowl. occ.	0.2607*	0.2085*	0.2319*	0.1878*
School-age children	-0.4150*	-0.3304*	-0.3822	-0.3359*
Commuting	1.063*	1.260*	1.134*	1.275*
Occupation:				
Computer specialist	-0.0033	-0.0052	0.0228	0.0183
Engineer	-0.3782*	-0.1694	-0.3470*	-0.1854
Life and phys. scientist	0.4511*	0.5649*	0.4513*	0.5990*
Regional characteristics:				
Knowledge-intensity:				
Employment share	0.1655	-	2.803	-
Home prices	-0.149*	-	-0.114*	-
Interregional access.	-1.342*	-	-1.703*	-
Economic conditions:				
Employment growth	-0.2528*	-	-0.2831*	-
Per-capita income	0.5823	-	0.6569	-
City scale				
Population	-0.3451*	-	-0.2926	-
Dummy:				
Stockholm	2.293*	-	1.964	-
Uppsala	0.1284	-	0.0194	-
Linköping	0.0431	-	0.3227	-
Lund	0.1611	-	0.0937	-
Göteborg	0.6697	-	0.5706	-
Umeå	0.1856	-	0.1643	-
Luleå	0.5202*	-	0.5434*	-
Likelihood	-5773.7	-5995.3	-4397.4	-4498.2
Rho-square	0.607	0.592	0.659	0.651

comparable to Herzog and Schlottman, since our dummy variables refer to very aggregate occupational groups.

As to regional characteristics, it can be noted, as expected, that 1980-1985 employment growth significantly retards out-migration, that high inter-regional transportation accessibility reduces the probability of migration as does the size of the labour market region.

While Herzog and Schlottman found that high-tech workers were less likely to relocate in the face of *lower* home prices, our results indicate that *higher* home prices reduce the probability to migrate. Our result seems reasonable, at least at this regional level, since the level of home prices is another reflection of economic growth and expansion.

As in the study by Herzog and Schlottman, certain regional characteristics are insignificant determinants of high-tech worker mobility. Among the significant variables is the somewhat odd result that the probability to out-migrate is higher for high-tech workers in Stockholm than elsewhere, although this coefficient is no longer significant when entrants to the labour market are excluded.

5.5 SUMMARY AND CONCLUSIONS

The growing knowledge orientation has made the availability of skilled labour and R & D resources a more decisive location factor. As a consequence migration and residential location among people in knowledge-related occupations has gained increased attention.

Our empirical study of migration among high-tech workers in Sweden replicates a US study. The Swedish study indicates a higher mobility among people in knowledge-oriented occupations than those in other occupations. Both studies demonstrate that life-cycle characteristics have a dominating influence on the decision whether to migrate to another region or not. The influence of the various characteristics is remarkably similar in both studies.

Both studies also show that different regional attributes and amenities have a significant influence on the migration decision. A straightforward conclusion is that a region which is more attractive for people with high-tech occupations will also be more attractive for high-tech firms.

However, more detailed studies are needed in order to list policy measures that could be undertaken in order to increase regional attractivity. It is not enough to model migration as a binary decision. The model should reflect the fact that people probably decide simultaneously whether to move and where to move. One possible approach for handling this would of course be to use a two-stage nested logit model. A related draw-back of our model application is the poor characterization of the regions. Surely, a much more extensive classification is needed in order to capture the pros and cons of different regions. Finally, a natural next step would be to apply a 'micro' approach, that is modelling the choice of migration and intra-regional location.

REFERENCES

Andersson, Å.E., 1986, "Presidential address: The four logistical revolutions", *Papers of the Regional Science Association*, 59:1-12.

Andersson, Å.E., Anderstig, C., and B. Hårsman, 1990, "Knowledge and communications infrastructure and regional economic change", *Regional Science and Urban Economics*, 20:359-376.

Andersson, Å.E. 1988, *Universities and the Future of Regions*, The Stockholm Office of Regional Planning and Urban Transportation, Stockholm (in Swedish).

Anderstig, C., and B. Hårsman, 1986a, "On occupation structure and location patterns in the Stockholm region", *Regional Science and Urban Economics*, 16:97-122.

Anderstig, C. and B. Hårsman, 1986b, "Why does Stockholm grow?", *PLAN*, 40:117-123 (in Swedish).

Greenwood, M.J., 1985, "Human migration: Theory, models and empirical studies", *Journal of Regional Science*, 25:521-543.

Herzog, H.W. Jr., Schlottman, A.M., and D.L. Johnson, 1986, "High-technology jobs and worker mobility", *Journal of Regional Science*, 26:445-459.

Herzog, H.W. Jr., and A.M. Schlottman, 1989, "High-Technology Location and Worker Mobility in the U.S.", in Å.E. Andersson, D.F. Batten and C. Karlsson (eds.), *Knowledge and Industrial Organization*, Springer-Verlag, Heidelberg, pp.219-232.

Jaffe, A.B., 1989, "Real effects of academic research", *The American Economic Review*, 79:957-970.

Klaassen, L.H., 1987, "The future of the larger European towns", *Urban Studies*, 24:251-257.

Premus, R., 1982, *Location of High Technology Firms and Regional Economic Development*, G.P.O., Washington, D.C.

Quigley, J.M., 1989, "Housing Allocation and Metropolitan Development", in *Urban Challenges*, SOU 1990:33, Swedish Commission on Metropolitan Problems, Stockholm.

CHAPTER 6

Some Consequences of 'Recurrence and Expansion': What Can We Learn about Logistic Networks from Chaos?

Donald G. Saari

6.1 INTRODUCTION

Understanding a logistic network involves developing an appreciation for the complexities associated with the component parts of a network. This includes an understanding of the movement of money, of commodities, of people, and of the production and consumption of goods and services. (See, for instance, Andersson and Johansson, 1984). The essence of many of these component parts can be captured by seemingly simple models. However, even simple models admit consequences that are surprisingly complex! Paradoxes abound, dynamical motion that intuitively should be 'simple' can be highly erratic, and small changes in assumptions can cause radically different conclusions. At first glance, this complexity is surprising if only because much of economics and decision theory is based upon concepts of aggregation and optimization - concepts that one might expect to introduce stability, reliability, and predictability to the conclusions of the resulting model. Yet, this need not be the case. Why?

In this Chapter I shed light on a major source of the erratic consequences of allocation, decision and economic models. I show that a basic and unifying explanation emerges if one modifies those ideas from the modern theory of dynamical systems that explain 'chaos' (See Saari, 1991). However, rather than a technical exposition, I will provide an intuitive approach emphasizing when and why one should anticipate the existence of such erratic behavior. Many of the concepts described here are introduced and illustrated with examples from decision analysis, statistics, and allocation systems. The technical material is described and motivated with informal arguments based on common examples such as the action of a bouncing ball. In this way I show how many of the unexpected outcomes associated with logistic networks can be understood and anticipated. In particular, I show that in any situation combining *expansion* and *recurrence*, one must investigate whether the accompanying behavior is erratic rather than predictable.

Situations that combine expansion and recurrence are common to the social and decision sciences. Of the two effects, recurrence is the more familiar as it is evident in business and other kinds of cycles. On the other hand, expansion effects are manifested through changes in discount or interest rates, inflation, and other standard parameters. In fact, expansion and recurrence can even arise as unintended, but consequential and accompanying aspects of the modeling. For instance, in price models for exchange economies with only two commodities, chaotic behavior *always* must occur if the system has three or more price equilibria; the appropriate recurrence and expansion effects must necessarily accompany the system with this number of equilibria (Chen, 1989)! So, whatever the reasons, one must expect expansion and recurrence to be combined in many, if not most models describing the component parts of logistic networks.

6.2 BOUNCING BALLS AND STABILITY

Why should one believe that *expansion* and *recurrence* can create highly erratic outcomes? Intuition for this is gained by considering the action of a bouncing ball. The motion of a ball dropped to a flat surface is quite predictable; indeed, even a child knows where a dropped ball will bounce. Much of athletics is based on this predictable behavior - a basketball player dribbles rapidly down the court without concentrating on the ball, a tennis player adjusts her racket to return the tennis ball to a desired position, and a golfer knows that only microscopic differences in his swing can be blamed for that slice.

The predictability of the motion changes should the bouncing ball be dropped on a curved surface; what happens depends on the curvature. For example, dropping a ball into a bowl yields a highly predictable motion; the ball bounces until it comes to rest at the bottom. Here the convex shape of the bowl introduces a *contraction* effect where the points of impact on the bowl become closer and closer to one another and approach the fixed stability point. It is this contraction effect with its induced stability that explains why ice hockey players prefer curved blades on their sticks and why many toys based on the throwing and catching of balls are designed to have curved pockets and receptacles.

The key fact about the convex surface that admits extensions of this notion to economics, physics, astronomy, the decision sciences, engineering, and other areas is the *contraction effect with its induced stability*. Quite correctly, stability is associated with predictability; in stable settings, slight errors and changes in initial starting positions, slight changes in assumptions introduce minor, if any, long term consequences. A hanging pendulum at the 6 o'clock position illustrates a most stable physical system; should this equilibrium position be disturbed, then very shortly the swinging pendulum returns to the hanging position.

Because stable motion tends to a final position, it is a concept that is comparatively easy to understand and to compute. This is because we can ignore the complexities of the actual motion by characterizing the eventual outcome - the final equilibrium position. For instance, rather than studying the actual peculiarities of a swinging pendulum, we characterize the pendulum's final static positions. It is this approach of emphasizing the final equilibrium positions rather than the actual dynamics that has proved to be so fruitful for economics, decision analysis, as well as aspects of engineering, physics, and other areas. This tractable approach has led to considerable success with theoretical and practical applications, so it is understandable why stability has dominated our thinking for centuries. In these notes, I emphasize 'instability' with its inherent loss of predictability. This kind of motion also can be illustrated with a bouncing ball, but now the curved surface is concave. With a concave surface the ball can rebound into any one of many different directions; a frustrating fact apparent to anyone attempting a game of tennis with a warped racket.

To appreciate the dynamics, consider dropping a ball on an inverted bowl. Well behaved motion results only should the ball be dropped with infinite precision directly over the top of the inverted bowl. Here the ball bounces along a fixed vertical line over the same fixed point. This fixed point at the top of the bowl is an *unstable equilibrium point*; it has properties similar to the precarious position of a pendulum standing upright at the 12 o'clock position. This equilibrium position is unstable because should we be unable to attain the required infinite precision of dropping the ball in the correct fashion, then the slightest error means that the actual motion will not resemble, in any manner, the bouncing action on the vertical line. Instead of mimicking the intended equilibrium behavior, the ball will hit a curved portion of the surface and rapidly bounce off the bowl. Notice that these points of contact of the bouncing ball on the bowl necessarily expand from one another; *this expansion effect characterizes the resulting instability.*

The expansion effect, created by the curvature of the bowl, introduces a deviation between the motion of the precise, theoretical orbit and the slightly altered, more realistic motion. It is this expansion effect that forces the difference between the motions of two dropped balls to become very large even should the initial starting positions be very close to one another. It is this expansion effect that forces all nearby motion away from the unstable equilibrium position. It is this dynamic of expansion that leads to a commonly used argument that such motion - the unstable equilibria - is highly unlikely to be observed, and so it is of little importance.

It is interesting to wonder whether this expansion effect can be harnessed in any manner. It can when accompanied by a recurrence effect. *Recurrence* is any effect that permits or forces the motion to return close to an earlier position. One can conceive of many scenarios which force the bouncing ball to return - for instance, place a second inverted bowl next to the original one. With two bowls, there are two unstable regions, and one can envision how these regions of instability introduce an interactive motion between each other. This interaction admits many different kinds of scenarios; the ball may bounce on each bowl once before rebounding to hit the other bowl; it may hit each bowl several times before rebounding to hit the other bowl; it may bounce off one bowl to hit a third one; or it might bounce off all of the bowls to enter some other region. In other words, it is this combination of *expansion* and *recurrence* that can create a complicated list of qualitatively different kinds of highly erratic behavior.

With recurrence the motion can return to a neighborhood of an unstable equilibrium. Thus a recurrence effect allows these regions of instability to regain their physical and practical importance. It is, then, the combination of these two traits of expansion and recurrence that create interesting and potentially important behavior that must be understood. The bouncing ball analogy illustrates why and how the combination of recurrence and expansion can create highly unpredictable behavior. Expansion takes points starting near one another, and expands their differences so that the subsequent motion quickly becomes very different. Recurrence has the impact of returning the motion to the vicinity of some original position. But, even should the motion return to the vicinity of an original point, the expansion effect permits the subsequent behavior to be strikingly different from what it was originally. With a continual return to the region, caused by the recurrence, one can envision the many, radically different kinds of possible behavior. Thus the expansion can be thought of as the factor forcing the diversity, the differences in future behavior. The recurrence can be thought of as the factor creating a continuing number of opportunities for the motion to return close to an original position in order to sample the diversity of behavior admitted by the expansion effect.

6.3 CYCLES AND EXPANSION

With our understanding of effects of expansion and recurrence, we must interpret accepted phenomena, such as the standard cycles from economics, in a new light. (Benhabib and Day 1982). This discussion suggests, for example, that even if an economic cycle returns an economy close to a previous position, this need not imply we are destined to repeat history. The expansion effects can create a radically different future. To illustrate, recall that thoughout this century there has been much discussion about business and other cycles - would they repeat some past performance, or would there be differences? With the coappearance of cycles (the recurrence effect) and expansion, it is clear that almost anything could happen; the future could be well-behaved by mimicking the past, or it could allow serious differences to occur.

In understanding the effects of recurrence and expansion, it is important not to attach value judgements to an assertion that the behavior is in a regime of 'unpredictable behavior'. There are settings where unpredictable behavior may be associated with the positive traits of introducing new opportunities, encouraging creativity, and admitting new situations that break from the cycles of the past. Then, there are settings where such an assertion is neither positive nor negative; but the conclusions serve as a valuable warning of what to expect. For instance, it may be that an optimal investment plan involves situations that combine recurrence and expansion. As we now know, this makes it possible for the optimal plan to be 'chaotic'. This knowledge is a valuable piece of information; it is an important warning that slight errors in the implementation of the plan could lead to significantly different outcomes; one must pay greater attention to the implementation of the plan during those periods of expansion. Finally, there are settings where 'unpredictable behavior' denotes that the desired effects will not occur. These different kinds of effects have been illustrated in theoretical models; some of these results are outlined below.

To start, recall that problems associated with the necessary trade-offs between production and present consumption to achieve growth have been effectively modeled through optimal growth and overlapping generation models. Once the models are described, the trade-offs between production and consumption are determined through optimization techniques. The resulting optimal strategy is called the *policy path.* Typical variables include the discount rate, the level of technology, and the utility function. As described in mathematical terms in a 1983 lecture (Saari, 1986) at the IMA (the Institute for Mathematics and its Applications), changes in the level of technology and the utility function can cause a recurrence effect; an effect where one must expect the policy path to return to a vicinity of an earlier position. This is intuitive - one can conceive of many circumstances where, say, to compensate for an over expansion we need a period of retraction that returns us close to an original setting. This is not uncommon in business. As also described in Saari (1986) (but in mathematical, technical terms) expansion effects can be introduced through changes in the discount rate, impatience, etc. This, too, is intuitive. Placing considerable emphasis on the future is a conservative act, one manifested by emphasizing future production over current consumption, so one should expect a concomitant form of 'stability'. This happens. On the other hand, placing emphasis on the present over the future significantly expands current consumption; this is a form of expansion. Thus, one must anticipate the policy functions in such settings to be chaotic in nature. Indeed, in 1985 two of the IMA participants provided details to prove that such motion does occur (Deneckere and Pelikan, 1986). Independently and around the same time, a stronger conclusion was obtained by Boldrin and Montrucchio, (1986).

What we have learned from this kind of analysis of optimal growth models is that even with optimization - a process often associated with introducing stabilizing effects - highly unpredictable, chaotic motion can and does occur! As illustrated by this discussion, the expansion effect can arise from unexpected sources - through changes in simple parameters of the system. In a more realistic setting divorced from the privileged simplicity offered by a theoretical model, one must expect that even more variables significantly influence the expansion factor - these can assume the form of inflation, interest rates, unemployment, changes in monetary policy, etc., etc. This illustrates that chaotic motion can be an unexpected consequence of even careful planning and that chaotic motion can be associated with 'positive activities'. This also illustrates that although long term planning is a necessity, short term monitoring and continual adjustments of resulting policy are necessary and required - particularly in a period of expansion and recurrence. In addition, this also demonstrates that chaotic motion must be anticipated in those many settings that, in a natural fashion, combine the traits of expansion and recurrence. This does not imply that such behavior necessarily will occur, but it may. (See the papers of R. Day, e.g., Day (1982) and Benhabib & Day (1982), for many other examples where these forms of behavior occur in economic models.)

On a different level, we now are learning that the notions of chaotic behavior are providing valued insights into what can, and what cannot be accomplished with certain approaches and programs. To illustrate, recall the supporting paradigm of free enterprise given by the *invisible hand* story of Adam Smith. Some may accept as a universal truth that the market forces of supply and demand will adjust prices to an equilibrium price where the markets clear. This may or may not be true, but, as of now, we do not know of any universal economic adjustment mechanism that supports this assertion. Even as recently as the early 1970's it was accepted that at least in simple 'text-book' economies of exchange, the price adjustment story, tatonnement, would force the market to adjust to equilibrium. This hope was dashed by an example constructed by H. Scarf (1960) proving that in economies with only three commodities, the standard price adjustment process never needs to approach equilibrium. What completely dampened any further hope for the existence of simple, standard procedures to work was research initiated by H. Sonnenschein (1972), advanced by R. Mantel (1974), and completed by G. Debreu (1974) asserting that the forces between aggregated supply and demand could be anything! Thus there is no reason to accept the standard story. In fact, using the Sonnenschein-Mantel-Debreu (SMD) conclusion, Saari and Simon (1978) proved that any theory advancing the idea of attaining equilibrium would require an unrealistic informational requirement.

How does chaos fit into this illustration? There are many studies advocating the use of various types of forecasting and the current and past behavior of the aggregate excess demand function in order to determine how prices should adjust in order to reach equilibrium. There are settings where these theories apply quite well, but can such a theory be modified to become universal? More generally, does there exist an universal theory that requires only a finite amount of information? Such an investigation, resulting in a negative conclusion, was carried out in Saari (1985). It was shown that even if one considers a very large, general class of price adjustment mechanisms involving current and past values of the excess demand function and the changes in this function, then, for any mechanism from this class, robust examples of economies can be constructed whereby the mechanism refuses to converge to equilibrium for robust sets of initial prices!

To understand this result, we need to appreciate those aspects of the economy which introduce the expansion and recurrence effects. The recurrence effect comes from 'over shooting'. No matter what the mechanism, Saari (1985) shows that there are choices of economies where the mechanism will 'over shoot' the equilibrium price. Namely, the market pressures of supply and demand will over inflate prices for certain goods and depress prices for others causing the new price to jump over the equilibrium price. Similarly, the expansion effect is created by over heated demand or an over abundance of supply for certain goods. This change in the excess demand has the same effect as the curvature of the bowl - it magnifies the resulting differential between subsequent prices created by a small, initial price difference. In this manner it was established that no universal theory exists that demonstrates how the market pressures of supply and demand will lead to equilibrium prices. (Incidentally, the same kind of ideas can be extended to show the difficulty in creating universal mathematical techniques of the kind often used in industry, optimization processes, etc., where the problem reduces to finding zeros of certain functions. (See Saari, 1987a.)

If there are no universal price mechanisms, then, maybe, the market mechanism needs to be adjusted to local situations. After all, one way to interpret the conclusion from Saari (1985) is that local customs and values, as reflected through the choice of utility functions and the resulting demand functions, strongly influence the kind of price adjustment mechanism that will, or will not work. This suggests that different locales may require the use of different kinds of mechanisms and that one should not expect that just because a procedure proves to be effective in one region, it will be effective in another. It also suggests that rather than adopting or searching for a global, universal mechanism, one should find those adjustment mechanisms peculiar to local conditions. The theoretic justification for such comments is found in Saari & Williams (1986).

A common message emerges from the consequences of these varied theoretic models. Namely, we must expect many of the standard components of a logistic network to be of the type that combines the factors of expansion and recurrence. Consequently, we must expect surprises. These surprises may be manifested by new opportunities, or by serious setbacks. In any case, attention and wariness is demanded.

6.4 DECISION PROCEDURES

One might hope to find refuge from such unpredictable behavior by retreating to the world of static problems. After all, the static problems from decision analysis and statistics do not involve dynamics, so one might expect an immunity from the unexpected outcomes associated with expansion and recurrence. Unfortunately, such immunity does not exist. As I demonstrate next, very closely related subtle mathematical factors can create unexpected outcomes. (Also, see Saari, 1987b). As just one consequence for, say, decision theory, it follows that there exist situations where, unintentionally, a decision maker can adopt decidedly inferior alternatives.

Consider a multi-attribute decision problem; say, trying to decide where to locate a new plant from among the three possible locations of A, B, C. In reaching this decision, suppose that fifteen equally weighted factors play important roles. For example, we might rank the three sites according to tax-related advantages, the availability of workers, transportation costs, etc. After this evaluation, suppose

- for 6 of the attributes, the rankings are $A \succ B \succ C$,
- for 5 of the factors, the rankings are $B \succ C \succ A$, and
- for the remaining four attributes, the site rankings are $C \succ B \succ A$.

A reasonable way in which to analyze this decision problem is to note that A is top-ranked on 6 attributes, B is top-ranked on 5 factors, and C is top-ranked on the remaining 4. From this information, it seems reasonable to rank the sites as

$$A \succ B \succ C; \tag{1}$$

an outcome that strongly suggests one should attempt to acquire the top-ranked location of A. Suppose now that, for some reason, site A is not available. The reasonable next step is to attempt to acquire B, the second ranked site. But, is B the next choice? In fact, C is ranked above B on 10 of the 15 attributes; quite clearly, $C \succ B$ so C should be selected! In fact, it turns out that when comparisons of the sites are made on a pairwise basis, the rankings are

$$C \succ B \text{ by } 10:5, \quad C \succ A \text{ by } 9:6, \quad \text{and } B \succ A \text{ by } 9:6. \tag{2}$$

These rankings from equation (2) are the exact opposite of those provided in equation (1). From these last rankings, a strong argument can be advanced that the originally selected A is the decidedly inferior choice, that the 'correct' ranking of the sites is $C \succ B \succ A$, and that C, not A, is the optimal choice!

As an illustration of what can go wrong with policy decisions based on statistics, suppose the unemployment rates of two cities each with a population of 30,000 are being compared. Suppose it is known that for men under 35 years of age, City I has a higher unemployment rate than City II, and, similarly, for men 35 and older, City I also has a higher unemployment rate than City II. Does this imply that, overall, City I has a higher unemployment rate than City II?. No. For instance, unemployment data asserting that

> for men under 35 City I - $9{,}000/24{,}000 > 1/3$, and for City II - $2{,}000/6{,}000 = 1/3$, while for men 35 and above City I - $3{,}000/6{,}000 = 1/2$ and for City II - $11{,}000/24{,}000 < 1/2$ satisfies the assertions.

But, the overall unemployment rates are $\frac{9{,}000+3{,}000}{24{,}000+6{,}000} = \frac{12}{30}$ for City I, with the higher unemployment rate for City II of $\frac{2{,}000+11{,}000}{6{,}000+24{,}000} = \frac{13}{30}$. These two examples illustrate the fact that one must take great care when interpreting rankings determined by statistical processes, decision procedures, and even voting outcomes. Somewhat surprisingly, static procedures of these kinds can be understood by introducing only a slight modification of the 'expansion - recurrence' analysis given above. To explain what type of modifications are necessary, note in the above analysis that the expansion factor played the role of allowing near-by points to have significantly different outcomes. For static problems, this role is assumed by the multiple kinds of heterogeneous profiles.

To see the role of heterogeneous profiles, notice that there are many different possible ways the data for the different sites could have occurred so that the outcome could have been $A \succ B \succ C$; an obvious one is that if for all fifteen traits the ranking had been $A \succ B \succ C$. There are many different kinds of profiles, that is, different listings of rankings for different traits, that lead (with the indicated method) to the ranking $A \succ B \succ C$. However, not all of these profiles lead to the same rankings for other sets of alternatives. Consequently, in order to determine whether there is a parallel for the expansion effect from the dynamics, the task is not just to determine what kinds of different profiles lead to the same ranking of the set of all alternatives, but also to different rankings of the various subsets of alternatives.

This task of finding a parallel for expansion has been completed. We now know that the greater the heterogeneity of the profile (e.g., the greater the differences among how the three sites are ranked over the fifteen attributes), then the greater the chance that there will be a 'decision paradox'. Thus, in decision problems, in voting situations, in statistical analysis and ranking, we should identify the possibility of 'heterogeneous profiles' with the 'expansion' property from the previous discussion about dynamics.

The final trait is one of recurrence. 'Recurrence' ensures that the various possible outcomes all are admitted. In static problems, this same trait is admitted should there be a large abundance of different kinds of profiles. But this can be seen to occur from simple combinatorics because with n alternatives (candidates, etc.) and 4 traits (voters, etc.) there are $(n!)^k$ different profiles - a number that increases in value with impressive rapidity! In other words, once the number of admitted alternatives exceeds two and once the number of attributes, voters, etc. exceeds or equals two, one must expect that the static system has inherited properties very similar to the 'recurrence' effect from dynamics. Indeed, these parallels have been used to explain and extend Arrow's Impossibility Theorem.

With these analogues and by mimicking the basic ideas from the analysis of chaos from the dynamics of expansion and recurrence, one can show that decision processes, processes of understanding demand characteristics of consumers and producers, etc., can be highly erratic. Perhaps the best way to provide support for these assertions is to give some examples of the types of conclusions that can occur. In the following recall that the plurality ranking of candidates is the process equivalent to the above multi-attribute process - it is a situation where each voter votes only for his top-ranked candidate, and if a candidate receives the most votes, then she wins.

THEOREM (SAARI 1989, 1990). *Suppose there are $n \geq 3$ candidates. This means there are $2n - (n+1)$ subsets of at least two candidates. For each of these subsets, arbitrarily choose a ranking of the candidates. There exists a profile of voters so that their plurality election ranking of each subset of candidates is the chosen one. There is only one way to tally each voter's ballot in order to minimize the number and kind of paradoxes that can occur. This is if the Borda Count (the BC) is used. This is where for k candidates, $k - i$ points are assigned to a voter's ith ranked candidate, $(i = 1, \cdots, k)$.*

An immediate consequence of this theorem is that one should not and cannot interpret election outcomes in a simplistic manner. The election outcome of $c_1 \succ c_2 \succ c_3 \succ c_4$ does not imply that candidate c_1 would be top ranked when compared with any other selection of these candidates. Indeed, unless the BC is used, it could be that c_1 is bottom ranked when compared with any other subset of candidates. Thus, one must treat the outcomes of elections, of decision processes, etc. with great care - they may not really mean what one thinks they mean.

Although the above assertion concerns voting, it applies equally well to multi-attribute decision making. The difference is that the alternatives play the role of candidates, and the different traits play the role of voters. To illustrate the second part of the above assertion, consider what would have happened to the location problem had the BC been used to determine the outcome. In this setting the ranking of the sites would have been

$$C \succ B \succ A \quad \text{with the tally} \quad 19:14:12; \tag{3}$$

notice that this ranking directly opposes the plurality outcome in equation 1 and that it is totally consistent with the rankings of the pairs in equation 2. As proved in Saari (1990), this is no accident; this is one of the basic properties that holds only for the BC. In fact, the BC turns out to be the unique way to tally elections and to determine the outcomes of multi-attribute decision problems in order to minimize the number, the kind, (Saari, 1990) as well as the likelihood of undesirable outcomes.

For non-parametric statistics, it has only recently been shown that the analogue of the BC is the well-known Kruskal-Wallis test (Haunsperger, 1991). This is the test that uniquely minimizes the number and kinds of unexpected, undesired statistical rankings for the different subsets of alternatives. Actually, statistical procedures admit far more diverse kinds of 'paradoxes' from the expansion and recurrence effect than do voting systems.

If the above general assertion holds true; if 'expansion' should be associated with the diversity of different ways individuals can alter their preferences about alternatives and if 'recurrence' can be identified with the abundance of different alternatives and profiles, then we must expect that many standard allocation procedures - procedures serving as component parts of a logistic network - admit outcomes with a similar kind of erratic outcome. This means that it can be dangerous and misleading to interpret outcomes from such procedures in a simplistic, natural fashion. This is, indeed, the situation. To illustrate, consider a simple problem of consumer preferences. Here there is an uncountable number of different preference rankings for each consumer; thus, if the above general assertion is to hold true, then we should expect manifestations of these preferences - say the excess demand function - to exhibit highly counter-intuitive behavior. That this is true is the conclusion of the Sonnenshein - Mantel - Debreu (SMD) theory mentioned above.

As it turns out, this SMD theory only serves as the tip of a huge iceberg. A more general assertion, similar to the 'voting theorem' above, also holds for the price mechanism. (See Saari, 1992.) For each of the $2^n - (n+1)$ subsets of two or more commodities, arbitrarily select a function satisfying the two basic properties of an aggregate excess demand function. Namely, it is homogeneous of degree zero (which asserts that it is the relative prices of goods that matter, not the actual prices) and it satisfies Walras' law (which is a budget constraint asserting that the money gained by selling goods has to be equal to or exceed the amount spent to purchase the items). One of the basic results in Saari (1992) is that there exists an exchange economy (initial endowments of wealth and preferences for each agent) so that the excess demand for each specified subset of commodities is the true excess demand function for this economy for most choices of prices.

In more common terms what this last conclusion means is that how prices change in the marketplace subject to market pressures need not have anything to do with how the prices will change should new goods be introduced, or other commodities be removed from the market. Information about the prices when one set of commodities is exchanged may give us absolutely no information about what would occur should a different set of commodities be exchanged. For one set of commodities, the price dynamics could be stable and well-behaved with a single price equilibrium; for another set of commodities, it could be unstable with many different price equilibria. As with the conclusions about voting and statistics, as well as with the chaotic motions of some systems, *almost anything can happen!*

6.5 CONCLUSIONS

In times of growth, in times of change, in times of innovation, in times of renewal and in times of creativity one must expect that the associated actions can combine the properties of recurrence and expansion. When this happens, when these twin traits occur together, one must expect that the unexpected can occur. This emergence of the unexpected may ensure new opportunities, or create unforseen difficulties. In any case, such situations demand constant monitoring to ensure that the eventual outcome is a desirable one.

Decisions - whether from standard decision analysis, voting, or statistical methods - admit closely related behavior. Here the signals indicating the potential of obtaining a 'recurrence and expansion' effect arise should the number of considered alternatives be greater than two, and should there be any degree of heterogeneity in the supporting data, voters' or consumers' preferences, etc. Such heterogeneity must be expected in times of change when opinions can differ greatly. In any such situation, the decision maker must be careful that the peculiarities of the system do not cause him to adopt inferior alternatives and that unjustified interpretations of surveys or of outcomes are not made.

REFERENCES

Andersson, Å. E. and B. Johansson, 1984, "Knowledge Intensity and Product Cycles in Metropolitan Regions", WP-84-13, IIASA, Laxenburg, Austria.

Benhabib, J. and R. Day, 1982, "A characterization of erratic dynamics in the overlapping generations model", *Journal of Economic Dynamic Control*, 4:37-55.

Boldrin, M. and L. Montrucchio, 1986, "On the Indeterminacy of Capital Accumulation Paths", in J.-M. Grandmont (ed.), *Nonlinear Economic Dynamics*, Academic Press, Boston, pp.26-39.

Chen, P., 1989, Northwestern University PhD dissertation.

Day, R. H., 1982, "The emergence of chaos from classical economic growth", *Quarterley Journal of Economics*, 98:201-212.

Debreu, G., 1974, "Excess demand functions", *Journal of Mathematical Economics*, 1:15-21.
Deneckere, R. and S. Pelikan, 1986, "Competitive Chaos", in J.-M. Grandmont (ed.), *Nonlinear Economic Dynamics*, Academic Press, Boston, pp. 13-25.
Haunsperger, D., 1991, PhD thesis, Northwestern University.
Mantel, R., 1974, "On the characterization of aggregate excess demand", *Journal of Economic theory*, 7:348-353.
Saari, D. G., 1985, "Iterative price mechanisms", *Econometrica,* 53: 1117-1133.
Saari, D. G., 1986, "Dynamical Systems and Mathematical Economics", in H. Sonnenschein (ed.), *Models of Economic Dynamics*, Lecture Notes in Economics and Mathematical Sciences, Springer-Verlag, pp. 1-25.
Saari, D. G., 1987a, "Informational requirements for the convergence of numerical methods", *Journal of Complexity*, 3:302-311.
Saari, D. G., 1987b, "The source of some paradoxes from social choice and statistics", *Journal of Economic Theory*, 41:1-22.
Saari, D. G., 1989, "A dictionary for voting paradoxes", *Journal of Economic Theory*, 48:443-475.
Saari, D. G., 1990, "The Borda Dictionary", *Social Choice and Welfare*, 7:279-317.
Saari, D. G., 1991, "Erratic Behavior in Economic Models", *Journal of Economic Behavior and Organization*, 16:3-35.
Saari, D. G., 1992, "The Aggregate Excess Demand Functions and Other Aggregation Procedures", *Economic Theory*, 2:359-388.
Saari, D. G. and C. Simon, 1978, "Effective price mechanisms", *Econometrica*, 46:1097-1125.
Saari, D. G. and S. R. Williams, 1986, "On the Local Convergence of Economic Mechanisms", in J.-M. Grandmont (ed.), *Nonlinear Economic Dynamics*, Academic Press, Boston, pp. 152-167.
Scarf, H., 1960, "Some examples of global instability of competitive equilibrium", *International Economics Review*, 1:157-172.
Sonnenschein, H., 1972, "Market excess demand functions", *Econometrica*, 40:549-563.

CHAPTER 7

The Complexity of Economic Decisions - Anticipatory Human Behavior

Wei-Bin Zhang

7.1 INTRODUCTION

During the past three decades, studies of nonlinear and unstable phenomena in various evolutionary systems have deepened our understanding of dynamic evolution (e.g., Haken, 1977, 1983; Prigogine and Stengers, 1984; Andersson, 1986, 1987; Zhang, 1988, 1989, 1990). Our thinking about evolutionary systems has been changed by the scientific concept of bifurcation. It has become clear that complicated forms of behavior such as chaos are not exceptions but rather universal phenomena in evolutionary systems. Rather than linearity, we have nonlinearity; rather than unique equilibrium, we have multiplé equilibria; rather than stability, we have instability; rather than predictability, we have unpredictability; rather than simplicity, we have complexity and chaos. Chaotic phenomena have caused great and even inflated interest among mathematicians, natural scientists and economists. But it is our wish that an understanding of chaotic evolution should not bring chaos into our thinking. The new vision about the evolution of dynamic systems should make it possible for us to systematically recognize the limitations and validity of classical scientific work. We have to organize and re-interpret existing knowledge in order to obtain new insights into the world.

This paper is concerned with nonlinear economic evolution. We examine the implications of synergetic economics and anticipatory human behavior. Synergetic economics was developed by Zhang (1991), on the basis of Haken's synergetics (Haken, 1977, 1983). It explores many of these issues within the framework of economic analyses. It examines the validity and the limitations of traditional comparative statics analysis. It also investigates the endogenous economic mechanisms which may result in complicated economic phenomena, such as regular and aperiodic oscillations and chaos. Furthermore, it deals with deterministic principles and stochastic processes in economic evolution. Haken's slaving principle and other mathematical methods shed light on economic dynamics with various adjustment speeds. The relationship between adjustment speeds and the length of the study period is thus a main concern of synergetic economics.

In order to examine the implications of synergetic economics at greater depth, this study will examine the interrelationships between individual behavior and the complexity of economic systems. We are particularly concerned with the anticipatory aspects of human behavior. Economic decisions are not only based upon past experiences, present conditions and rational knowledge. One's vision about the future also has significant effects upon decision making.

The Chapter is organized as follows. First, we examine the role of anticipations in decision making. Second, we examine the implications of synergetic economics for 'anticipatory phenomena'. Third, we examine economic planning from the viewpoint of synergetic economics. Finally, we present our conclusions.

7.2 ANTICIPATIONS IN DECISION MAKING

There is a great difference between social systems and physical ones. The behavior of the 'elementary particles' of social systems - human beings - is anticipatory in the sense that decision making is based upon the decision maker's vision about the future. There are always two worlds - the objective and the subjective - with which the decision maker is faced. He lives in the objective world, while the subjective world influences his perception and the way in which he makes his decisions. If the two worlds were always identical for every member of the society at any moment, then economic activity would become very simple - in this situation the picture presented in the classical Arrow-Debreu equilibrium model or Solow's growth model would be valid for the real world.

However, an agent's behavior is affected by past experiences, present conditions, rational knowledge and his vision about the future. Anticipation appears in every aspect of life. This can be seen by looking at the decision making processes of entrepreneurs, producers, consumers, bankers, stockbrokers and currency dealers.

As an example, we may look at the behavior of the profit maximization of a firm. The firm's short-run profit is given by

$$y = pF(K, L) - rK - wL,$$

where F is a production function, r the interest rate, w the wage. The firm has to decide how much labour L and capital K should be employed in order to get a maximum profit. For given p, r and w, under certain conditions the firm can uniquely determine L and K as functions of p, r and w. But in reality p, r and w may not be actual values. Instead, they may be what the firm imagines when the decision is made. For instance, the price of the commodity may be changed after the firm has finished production. In fact, when a firm makes decisions, it has to forecast many factors such as market situations, policies, the tax structure, technological progress and the like. The success of the firm is dependent upon its ability to forecast.

Expectations are a natural subject for an analysis of economic decisions, whose effects belong to the future. For example, a farmer may decide to plant corn today, but the corn is not harvested until later. Obviously, the farmer has to estimate or guess the price of corn. Most nineteenth century economists attempted to examine this problem in capital theory by using the idea of stationary states in which prices are always constant over the study period. But there is an exception. Alfred Marshall did attempt to deal with uncertainty and price expectations by two distinct expectation theories: one theory for the short run and another for the long run. Marshall's long-run theory can be illustrated by the following description of how a cloth manufacturer forms expectations about the future wages he will have to pay his workers. He notes that 'in estimating the wages required to call forth an adequate supply of labour to work a certain class of looms he might take the current wages ... or he might argue ... that looking forward over several years so as to allow for immigration he might take the normal rate of wages at a rather lower rate than that prevailing at the time ... or might think that wages of weavers were abnormally low in consequence of a too sanguine view having been taken of the prospects of the trade a generation ago' (Marshall, Chapter V, Section 1, 1920). By defining the normal price to be the price where the supply curve intersects the demand curve, Marshall's short-run analysis assumed that firms will not directly observe the factors that shift the demand curve. Instead, firms see that the price they get for their usual output changes. When this occurs, firms have static expectations; i.e., today they invest, expecting tomorrow's price to be the same as that of today. But this will lead to a new output tomorrow and a new market clearing price which will not be identical to the expected one.

Keynes (1936) distinguishes between short-term and long-term expectations. The distinction is quite fundamental in the sense that it cuts across the whole field of business behavior in such a way that two distinct theories can be formulated for either of them. The distinction runs along the same lines as the Marshallian distinction between the short run and the long run in the theory of the firm: the criterion is whether the capital equipment and the set-up of the firm are constant or may change. Short-term expectations are related to the proceeds and costs to be received and paid by firms when the process of production on which they embark is completed. The firms decide how to use certain types of capital equipment and organizational structures which are at their disposal, in the light of the expectations. Long-term expectations are related to the proceeds from, and the cost of, additional capital equipment. The subject of expectations is not related to strategies for a given type of capital equipment but rather the proceeds and costs of an optimum strategy of output from an increase in the volume of capital equipment.

Hicks (1939) suggested the term 'perfect foresight' for the situation where producers anticipate a price and then make production decisions. 'Perfect foresight' means that the actual price is identical to that anticipated. The key point that Hicks emphasized is that the expectations held by firms about the future are endogenous variables. That is, there is an interrelationship between visions about the future and actual events in the future.

It is not difficult to see that an actual form of behavior may depend on several expectations, each of them related to a different subject. For instance, an entrepreneur makes an estimate of the price at which he expects to sell his products and of the costs which he expects to incur. The output which he decides to produce depends then on both these expectations. We can classify expectations as primary and intermediate expectations. Primary expectations are those in which there are no other expectations between the evidence and the prediction. Intermediate expectations are derived from primary expectations. For instance, a businessman's expectations to gain large profits may be derived from his primary expectations of high prices and low costs.

Anticipation is also dependent upon the complexity of a system. In a relatively simple system with relatively simple dynamic behavior, the formation of expectations is relatively simple. For instance, in an agricultural economy, expectations about products may be very simple. The expectations may simply be formed according to past experiences. But in modern stock markets expectations tend to be more complicated.

Anticipations are cultural and are thus dependent upon the value system of the culture. The way people form their expectations is an important characteristic of the culture. Obviously, people in a free society have different ways of thinking about the future compared to those in tyrannies.

Anticipations are also influenced by institutional and organizational structures. The success of Japanese industrialization can hardly be understood if one is unaware of the way that Japanese workers and businessmen think about the future. This problem is, in fact, fundamentally related to incentive problems. The success of Japanese industrialization provides an example that Adam Smith's division of labour, with freedom of choice of working places, may not be an efficient economic system in terms of long-run economic development. We will explore this issue further in another context. On the other hand, the high degree of creativity of Europeans throughout history cannot be explained if one does not investigate how their institutions, habits and other cultural aspects affect the way that these people think about the future (surely, we can assume that each nation has a significant potential of genius). People can hardly realize their creative potential in a culture in which the production of new ideas always leads to punishment.

7.3 ANTICIPATION IN A CHAOTIC WORLD

Recent studies of nonlinear dynamic economic phenomena tell us that even if economic behavior is described by a very simple model, the behavior of the model may be complicated. Chaotic economic phenomena are universal rather than special characteristics of economic evolution. The idea that a model with a relatively simple structure may exhibit chaotic behavior has shocked scientists trained in traditional 'determinism'. It may be said that we are still far from an understanding of the implications of the message of chaos theory. We cannot expect to understand the behavior of an actual dynamic system accurately, even if we have exact knowledge about the rule controlling its behavior. The relationship between 'randomness' and order turns out to be more complicated than we have assumed.

I will not make detailed comments about the complexity of dynamic systems as the Chapters by Haken, Andersson and Saari provide excellent introductions to this topic. In addition, many examples of reasonable economic models exhibiting instability and complicated behavior are provided in Zhang (1991).

I will now emphasize some aspects of chaos before I deal with the main topics: (i) in this study chaos refers to the solution of differential and partial differential equations; in other words, chaos is a consequence of dynamic, deterministic principles; (ii) chaos is unpredictable but bounded; (iii) social chaos is different from 'anarchy'; (iv) social chaos does not deny the validity of traditional approaches; but traditional approaches can be considered as special cases; (v) although chaos is universal in economic evolution, this does not imply that chaos necessarily exists in any sector at any period; there are relatively stable sectors in chaotic economic systems; it is significant to study the relations between 'stable sectors' and 'chaotic markets'; (vi) chaos is a time-dependent phenomenon, but a progressive society has to face the problems created by chaos; (vii) chaos makes short-run and long-run effects of economic policy uncertain.

In the remainder of this Section, I will examine the relationship between chaos and anticipations.

The preceding Section mentioned that economists have been aware of the significance of the influence of expectations upon the economic system for a long time. Keynes' animal spirit, Hicks' distinction between short-run and long-run equilibrium as well as rational expectation macroeconomic models, Iwai's work on the Wicksellian cumulative process, and Grandmont's - and other economists' - short-run equilibrium models reflect the recognition of the anticipatory characteristics of human behavior. Many analytic models explicitly take certain expectation patterns into account. But it can hardly be concluded that fruitful results have been obtained from this analytical work. We do not yet understand the relationship between the complexity of a system and anticipatory behavior. In this Section, we will discuss the complexity of anticipatory behavior in the light of chaos theory.

We consider that instability and anticipation can be 'a pair of twins' in the sense that anticipation may become very simple if the system is stabilized; and the system may tend to be stabilized without anticipation. In terms of synergetic economics, anticipation tends to be very complicated if the system is located at critical points. This can be understood if one looks at the current politics of some socialist nations. The complexity of money markets and land markets also reflect the complicated interactions between expectations and the dynamics of the system.

In what follows, we will apply our general ideas to explain the characteristics of entrepreneurs. The concept of an entrepreneur plays a central role in the work of Schumpeter. It should be noted that according to the Arrow-Debreu equilibrium approach of neoclassical economics, one cannot explain the reasons for the existence of entrepreneurs. According to the fundamental hypotheses of presumed stability, it is natural to reach a logical conclusion concerning the nonexistence of entrepreneurship because in the long term no one can possibly obtain positive profits in a stabilized system.

In The Theory of Economic Development (1934), Schumpeter argues that development should be understood as only such changes in economic life that are not forced upon it from without, but rather arise by their own initiative, from within. He identifies the key development process as the 'carrying out of new combinations'. In a competitive economy new combinations mean the competitive elimination of old combinations. It is the entrepreneur who carries out the new combinations. The entrepreneur leads the means of production into new channels and may thereby reap an entrepreneurial profit. The entrepreneur' also leads in the sense that he draws other producers in his branch after him. But as they are his competitors, who first reduce and then annihilate his profit, this is, as it were, leadership against one's own will'.

Schumpeter's theory of the entrepreneur is part of the theory of capitalist economic development. As mentioned above, according to Schumpeter, entrepreneurs are economic agents whose function is the 'carrying out of new combinations', the creative destruction of equilibria, thereby preparing the ground for a superior state of equilibrium.

An essential aspect of Schumpeterian competition is the notion that there are winners and losers and that the process is one of continuing disequilibrium. Firms facing the same market signals respond differently, and more so if the signals are relatively novel. In order to obtain entrepreneurial profits, firms have to solve many technical, economic and psychological problems. Firms participate in innovative competition through product development, price or cost competition through process development, and marketing competition through sales or market promotions. Failures in any competitive field may affect the incomes of the firms.

According to Swoboda (1984) Schumpeter's entrepreneur can be described by the following five qualities:

(i) The entrepreneur can be - but need not be - the owner (main shareholder) of the firm. There are distinct types of entrepreneurs: the sole owner, the industrial leader (who still owns the majority of shares), the employed manager, and the promoter (founder), the purest type of entrepreneur. Schumpeter's conception of the financing of firms can be characterized by the neglect of equity financing, rather stressing financing by retained earnings and overemphasizing credit financing.

(ii) The entrepreneur is an economic leader, whose main task is not to invent or to create new possibilities, but to carry them out. In order to be successful, keenness and rigour, but also a certain degree of narrowness which concentrates on immediate chances are essential. Schumpeter stresses that in economic life decisions must be taken without working out all the details and he is by no means convinced that gathering and exploiting information is significant for the functioning of entrepreneurship. 'Thorough preparatory work, and special knowledge, breadth of intellectual understanding, talent for logical analysis, may under certain circumstances be sources of failures'.

(iii) The entrepreneur is never a risk bearer; he only bears risks if he is also a capital owner. This statement has often been subject to criticism. The activities of a pure manager can be quite risky if the variance of his human capital is considered to be a measure of risk.

(iv) The objective of the entrepreneur is not the maximization of the present value of his income or consumption. He works restlessly, not living to enjoy his acquisition, but rather being driven by 'the will to conquer', 'the impulse to fight', the 'joy of creating'.

(v) Entrepreneurial activity is not a factor of production.

There are other definitions of an entrepreneur, which are also discussed in Swoboda. According to Knight (1921), the essential quality of an entrepreneur is the bearing of uninsurable risk. It is endured because of the chance of profit. One of the main problems dealt with by Knight is whether the separation of control and guarantees (risk bearing) is possible. He concludes '...when control is accurately defined and located, the functions of making decisions and assuming the responsibility for their correctness will be found to be one and indivisible'. The natural result is a complicated division or diffusion of entrepreneurship, distributed in the typical modern business organization by a hierarchy of security issues carrying conceivable graduations and combinations of rights to control and to freedom from uncertainty as to income and vested capital.

Kirzner (1973) defines an entrepreneur as a decision-maker who is alert to hitherto unnoticed opportunities. For Schumpeter the entrepreneur destroys an equilibrium, while Kirzner's entrepreneur helps to attain an equilibrium by eliminating given possibilities of arbitrage.

Schultz (1975) defines entrepreneurship as 'the ability to deal with disequilibria' by reallocating resources, attaching the notion of entrepreneurship to workers, households and students, who respond to changes in expected earnings as a result of different studies. The ability to reallocate resources can only be studied adequately in equilibrium models. With respect to such models entrepreneurial ability is a scarce factor of production which can be subject to improvements by education. The reallocation of resources is never without risk.

According to Casson (1982), an entrepreneur is 'someone who specializes in taking judgmental decisions about the coordination of scarce resources'. 'A judgmental decision is one where different individuals, sharing the same objectives and acting under similar circumstances, would make different decisions'. They may have different access to information and so on. Casson's and Schumpeter's entrepreneur have only one common feature: the entrepreneur need not be the sole owner of the firm. Casson's entrepreneur need not demonstrate direct leadership or organizational skills if he is able to delegate. He only need show judgement. Entrepreneurial activity, which is motivated by self interest, is a factor of production with a market of its own, or several markets of its own. The entrepreneur bears some risk. Casson emphasizes the relationship between the entrepreneur and the information economy since better judgement is facilitated by better information. Casson sees that it may be difficult for an entrepreneur to raise funds in order to exploit his better judgement based on an information advantage. He further argues that the entrepreneur may not get funds from a private financier owner since the latter may not have the information held by the former. In this case, banks are needed. He thus explains the capital structure decisions of firms and the existence of banks.

It should be noted that there is only one common characteristic shared by Schumpeter's entrepreneur and the 'post-Schumpeterian' entrepreneur: the entrepreneur need not be the owner of the firm. For post-Schumpeterian theorists, the entrepreneur is an endogenous part of capitalism and is motivated by self interest with risk bearing. The post-Schumpeterian entrepreneur does not necessarily relate to the development of capitalism. Post-Schumpeterian theories do not advocate a decreasing importance of the entrepreneur in the course of the development of capitalism, and they do not restrict the role of the entrepreneur to the implementation of innovations. The entrepreneur generally helps to attain an equilibrium and does not disturb it.

We tend to accept the post-Schumpeterian concepts of entrepreneurship. For us, a successful entrepreneur is someone who has a special ability to anticipate the future, taking his own behavior as an endogenous variable of the whole system. A successful entrepreneur is a master at solving 'the future market equations' on the basis of his information, intuition and knowledge.

Entrepreneurs act in a rational but chaotic system. By a chaotic system, we mean that the system may exhibit chaos under certain conditions. There are relationships between entrepreneurship and chaos. We argue that if the economic system is stabilized, there will be no entrepreneurs in the long term. This is the reason why entrepreneurship cannot exist in the neoclassical economic system, which is assumed to be stable. If the system is stable, it will tend to stay at some equilibrium. In such a stabilized system, small changes in parameters can only cause small shifts of the variables. In other words, small changes in people's tastes or production technology can only result in small changes of profits. Hence, it is not difficult to see that entrepreneurs will not be active in a stabilized world since their ability as entrepreneurs can hardly help them to obtain excess benefits in the long term. This character of the stabilized system leads Schumpeter to conclude that in the very long term entrepreneurs' profits tend to be zero.

It is necessary to emphasize a few characteristics of our entrepreneur in a competitive economic system.

(i) The success of an entrepreneur is due to the failure of other participants. Winners and losers cannot be separated. Success or failure is due to interactional forces among entrepreneurs and consumers who have different viewpoints about the future.

(ii) In a complicated knowledge society, there are always opportunities to obtain excess profits. Such opportunities would exist even if technology was fixed and information was relatively well-known to everyone.

(iii) Even if he is smart enough, an entrepreneur cannot always succeed. Since the system is possibly chaotic, an entrepreneur cannot always correctly forecast the future market. But information, knowledge and intelligence tend to improve his opportunity to obtain excess profits from the exchange.

(iv) There are no deterministic relationships between education and the success of an entrepreneur. Good students with an education in economics do not necessarily become successful entrepreneurs. Appropriate education can improve the students' capacity to obtain opportunities, but it cannot guarantee success. This is not due to the failure of rational knowledge, but due to the characteristics of chaotic systems. Intuition, which helps one to recognize the 'future' situation of the system under consideration, often plays a significant role in obtaining excess profits.

(v) Entrepreneurs exist due to the potential instability of the economy. Thus, economic instability and entrepreneurs depend upon each other. If the system is stabilized or otherwise predictable, we can hardly expect the existence of entrepreneurs in the long term.

7.4 IMPLICATIONS FOR PLANNING

We have examined some interrelations between anticipations and the complexity of economic dynamics. We now examine the implications of our previous discussions for the economic planning of a government.

Economic planning is extremely complicated in a mixed economy where individuals have the freedom to choose their levels of consumption, savings and investment subject to the interventions of the government. The success or failure of governmental planning depends upon the planners' ability to forecast the behavior of the population in the future. On the other hand, planning can change the way that people anticipate the future. In modern democratic economic systems, the degree to which the government can successfully play the game depends upon the knowledge and visions of the future encompassed by its population.

In principle, no successful form of planning can be achieved without a profound forecasting system. However, in free systems, forecasting is almost impossible. This implies that the goals of much of economic planning can hardly be realized in practice in the long term. I do not think that it is possible to find any long-run plan (for instance, longer than 10 years) which has been implemented as originally planned. But in a chaotic world, such planning is often important because it plays a role in preventing catastrophes within systems and provides stimulus for a society which has not enough knowledge stocks to handle planning with genuine freedom. Through planning, we are guaranteed a certain suboptimal and viable solution rather than some form of uncertain universal optimization.

It should be noted that the explicit distinction between long-run and short-run analysis in synergetic economics is important for planning. The effects of many types of planning such as infrastructure development and education investment can only be reflected in the long term.

The emphasis on education made by various governments has an extreme impact upon the accumulation of knowledge within populations. The significance of creativity and knowledge can only be reflected in long-run economic development. Myopic viewpoints about the future often result in harmful decision-making by governments. Educational problems are especially serious in underdeveloped nations, where only short-run benefits appear to be of interest for the people. Such a harmful environment is often due to a misunderstanding of the government in relation to education and knowledge.

Governments have different options with respect to infrastructure, which is always of a long-run importance. They may only make investments in infrastructure when serious bottlenecks occur due to an expansion of regional or urban development. In this case, the government plays a passive but predictable role in regional development in the sense that infrastructure follows regional development with less likelyhood of investment profitability. The government may also supply infrastructure as an engine for regional or urban development. However, this strategy of investment in infrastructure is risky because the response of entrepreneurs is highly uncertain.

7.5 CONCLUSION

This study explores the complexity of economic evolution and anticipatory behavior in the light of synergetic economics. However, synergetic economics attempts to reveal many other aspects of social evolution. In what follows, we mention just a few directions for its potential applications.

Synergetic economics provides a new link to unite different ideas in economics and politics. It has also provided further insights into the order and disorder of social evolution. We can now systematically recognize the relations between and the limitations of economists such as Smith, Malthus, Ricardo, Marx, Walras, Schumpeter, Mises, Keynes, Simon and the modern competitive equilibrium school. Such a synthesis can be furthered by an understanding of the complexity of dynamic evolution. The modern concepts of short-run and long-run analysis, fast and slow adjustment, stability and instability, anticipation, knowledge, complexity and chaos can help us to deal with complicated nonlinear phenomena.

Although this Chapter is limited to making some reflections on economic problems, we should like to mention that the new vision about evolution could also play a role in explaining some other social problems. Not only have we deepened our thinking about social progress and the complexity of social evolution - areas in which so many distingnished men have devoted their time and energies. It also appears that a new appreciation of the characteristics of various cultures is going to be achieved. For instance, applying the concepts of stability versus instability and equilibrium versus chaos, we can derive some hints about the differences between the Western culture and (east) Asian Culture.

One could claim that the most important aspect of all cultures is individual happiness. So religions, ethics, laws, institutions and the like should be designed for the happiness of the population. In order to guarantee individual happiness, social order might be necessary. Each culture (or society) designs social institutions and develops friendships and habits in order to achieve social order. Besides the concept of order, the most important aspect of culture is the concept of stability. We may thus conclude that the fuzzy concepts of individual happiness and stability are the two most important concepts for analyzing cultures. New visions of evolution are significant for understanding cultures, because they have deepened our insight into the concept of stability.

There are many other essential issues to be examined. For instance, we do not know much about the interrelations between information, anticipation and the stability of economic systems. It is also important to examine the relationships between anticipations and incentives.

There is nothing new in recognizing the role of anticipations in human behavior. The most important factor, however, is to recognize their 'weight' in scientific analysis and to study them systematically. Modern economists have introduced anticipatory behavior into economic analysis. But unfortunately, the main emphasis is upon the price system without examining other aspects of anticipations. For instance, there are no systematic studies about anticipatory behavior in connection to the interrelationships between institutional structures, organization and incentives.

REFERENCES

Andersson, Å.E., 1986, "The four logistical revolutions", *Papers of Regional Science Association* 59:1-12.
Andersson, Å.E., 1987, "Creativity and Economic Dynamics Modelling", in D. Batten, J. Casti and B. Johansson (eds.), *Economic Evolution and Structural Adjustment*, Springer-Verlag, Berlin.
Casson, M., 1982, *The Entrepreneur*, Martin Robertson, Oxford.
Haken, H., 1977, *Synergetics: An Introduction*, Springer, New York, Berlin.
Haken, H., 1983, *Advanced Synergetics*, Springer, New York Berlin.
Hicks, J.R., 1939, *Valuc and Capital*, Oxford University Press, London.
Keynes, J.M., 1936, *The General Theory of Employment, Interest and Money*, Brace and Company, Inc., Harcourt.
Kirzner, I.M., 1973, *Competition and Entrepreneurship*, University of Chica-
go Press, Chicago.
Knight, F.H., 1921, *Risk, Uncertainty and Profits*, University of Chicago Press, Chicago.
Marshall, A., 1890, *Principles of Economics, 8th edn*, 1920; variorrum edition, ed. C. W. Guillebaud, 1961. Macmillan, London.
Prigogine, I. and I. Stengers, 1984, *Order out of Chaos: Man's Dialogue with Nature*, Bantam Books Inc., New York.
Schultz, I.W., 1975, "The value of the ability to deal with disequilibria", *The Journal of Economic Literature*, 13:827-846.
Schumpeter, J.A., 1934, *The Theory of Economic Development*, Harvard University Press, Cambridge, Mass.
Swoboda, P., 1984, "Schumpeter's Entrepreneur in Modern Economic Theory", in C. Seidl (ed.), *Lectures on Schumpeterian Economics*, Springer-Verlag, Berlin.
Zhang, W.B., 1988, "Hopf bifurcations in multisector models of optimal economic growth", *Economic Letters*, 26:329-334.
Zhang, W.B., 1989, "Oscillations in Rodriguez's model of entry and price dynamics, *Journal of Economics and Control*, 13:485-497.
Zhang, W.B., 1990, *Economic Dynamics - Growth and Development*, Springer-Verlag, Heidelberg.
Zhang, W.B., 1991, *Synergetic Economics*, Springer-Verlag, Heidelberg.

CHAPTER 8

Incomplete Information and Logistical Network Equilibria

Kiyoshi Kobayashi

8.1 INTRODUCTION

In his stimulating book 'Infinite in All Directions', F.J. Dyson manifested the possibility of new frontiers for the 'Science of Diversity', and presented a disputable hypothesis on the evolution of organisms (Dyson, 1988). According to his arguments, the most fundamental mechanism for catalyzing the evolutionary process should be sought in the fact that biological synthesis is ingeniously accomplished by interactions between the hardware and software of genes. The preservation of the species can be accomplished only by perfectly-functioning hardware. No error is allowed in the replication mechanism of genes. On the other hand, each species, surprisingly enough, has shown infinite diversity in its individuality (software) as can be easily demonstrated in humankind. Thus, the separation of the hardware and software of the organism enables every species to create immensely complicated systems and to retain the compatibility between the perfection in its replication process and the infinite dimensions in its diversified individuality.

Historically, humans have created numerous break-throughs in the hardware technology of social systems. It should be noted that hardware such as language and money play indispensable roles in our society. Money and languages can function perfectly within nations' economies. Networks, technology, and knowledge, etc., are also elements of fundamental hardware which facilitate the activities of our society and economy. As Å.E. Andersson stated, sometimes our world evolves dramatically with the advent of completely new functions and structures of social and economic systems (Andersson, 1986). These drastic forms of evolution which occur when humans create new breakthrough hardware technology and succeed in introducing new technology into our society are known as logistical revolutions.

History can easily show that there are two fundamental preconditions common to all logistical revolutions. The first precondition is that the newly emerged hardware technology be perfectly accepted by the societies and economies of the time. The knowledge and information about the new form of hardware becomes common knowledge among people and organizations. In other words, it serves as a major source of infrastructure for the societies. The second precondition is, more significantly, that the advent of new logistical systems to society increases the dimensions of human activities. Human activity patterns are enriched with increased flexibility and diversity. As time passes, the old types of logistical systems also evolve with a sequence of technological innovation. Thus, it is fair to say that as a whole logistical systems evolve and change to relax and remove restraints imposed on the actions of firms, agents and organizations, and to foster the emergence of new types of activities.

The fundamental properties of logistical systems can be summarized as follows: 1) a system is comprised of both hardware and software; 2) its hardware can be characterized by a certain 'perfectness', in the context of my arguments, in its functioning; 3) it can accept infinitely diversified software (programs); 4) it can distinguish correct programs from erroneous ones; 5) there should be synergetic interactions between anonymous users who simultaneously utilize hardware resources. The final remark refers to the fact that every agent might possess private information, characteristics and preferences which are not observable by others. Thus, logistical networks are designed to cater for synergetic interactions between large numbers of anonymous agents whose behaviour is led by decentralized decision making and information processing.

This Chapter presents a new analytical framework for logistical network equilibria with incomplete information. The logistical networks can be characterized by a set of hardware resources and a set of programs to be accepted by the hardware. A user of a logistical network should choose or develop an admissible program to combine the hardware resources with his inputs. Due to the limitation of resource availability, there should be interactions between its potential users. The basic element of our network equilibrium concept is differential information; different users have different information about the availability of software; they choose their programs on the basis of their private (differentiated) information. The purpose of this Chapter is to develop a general equilibria concept that makes explicit the information or knowledge that a user has as part of his primitive characteristics. The model we present is a reinterpretation of Harsanyi's model of incomplete information games. It differs from Harsanyi's approach in the explicit consideration of rational expectations formation by users. A numerical illustration attempts to provide pedagogical insights into logistical network equilibria with incomplete information.

8.2 BASIC CHARACTERISTICS OF LOGISTICAL NETWORKS

8.2.1 Characterization of Logistical Networks

A logistical network - briefly, a network - may be viewed as a type of complex of hardware and software. The hardware of a network accepts as inputs a basic set of natural and artificial resources as well as programs (software). The software designates the precise ways to produce outputs; it arranges the way in which the network hardware and resource inputs are combined; it regulates the processes which are coordinated to produce final outputs. Typically, the outputs of networks are the allocation of resources and commodities, the locomotion and movement of resources, the provision of knowledge and information, and the creation of products and services.

Thus, a logistical network can be characterized by its software Υ, hardware Π, and environment e and designated by $L = (\Upsilon, \Pi, e)$. We define the terms software and hardware in a much broader sense than that used in the field of information science. Hardware is a kind of 'mechanics' which is designed to synthesize different types of input resources, information, knowledge and energy in order to produce final outputs, e.g. commodities, services and knowledge. The hardware technology of a logistical network is assumed to be fully described by its software arrangement. We regard software as a set of rules which regulates the ways to synthesize input resources; which rules the modes of communication among network users; which catalyzes synergetic interactions among users; which designates the precise ways of material and immaterial processing.

Information about the hardware of networks is, in principle, available to the public, and forms a part of the common knowledge for all users in Aumann's sense (Aumann, 1976). The basic data of the hardware of networks consist of the dataset of resources space and network technology. The software Υ means the set of programs which are designed to combine hardware resources with inputs. Thus, every program can be characterized by the set of hardware resources (subsets of hardware space) to be used in its operation. The flexibility and availability of a logistical network is crucially influenced by the hardware technology and the restraints imposed on the availability of software. The profits and gains which a user can obtain through the usage of the networks are essentially dependent upon the other users' actions. The synergetic interactions among users form an essential part of the network availability of logistical systems. Logistical networks with the flexibility to accept diversified software serve as platforms to catalyze coherent and integrated behaviour.

The basic data of the economic environment consist of agents (users) $T = \{1, \cdots, N\}$, and the characteristics of each agent, which typically are, for agent t, his preference relation $\succeq_t$ (or its representation by a utility function U_t), his technology, given say, as a software possibility set, and his information system Λ_t. Denote the characteristics of the tth agent by (U_t, Λ_t) for $t \in T$. With these assumptions, a specified economic environment of logistical network e determines the N-tuple of characteristics of the N-tuple of agents. Note that this formulation does not exclude externality.

8.2.2 Information Space

Incomplete information in network equilibrium models includes uncertainty about agents' characteristics, preferences, and the whole body of knowledge about network software. A complete description of a network should resolve these uncertainties. If a state of nature is known and the agents' choices of software are known, each agent will know the outcome and his corresponding utility. In contrast to the single-person decision problem, we consider N agents. Thus, a complete description of a state of nature must contain information not only for resolving the uncertainties, but also for determining the extent to which each agent knows the state of nature. There must be some states of nature that are indistinguishable from others if there is incomplete information. The degree to which the natural states are indistinguishable will affect the agents' behaviour and must be part of the description of a natural state. Recognizing the possible ability of agents to differentiate among states allows us to analyze asymmetric information among agents. This is the essence of incomplete information beyond that treated by standard decision theory.

Let us explain how one can formally describe agents' information about agents' characteristics, preferences and choice space. Each agent has his own information about himself, $\omega_t \in \Omega_t$ which is not observable by all other agents. The space of information shared by agent t is denoted by Ω_t, which is assumed to be a separable metric space and is endowed with the corresponding Borel σ-field $A_t := \beta(\Omega_t)$.

Further, we assume that there exists a whole space of information Θ which is defined by a product space of each agents' information space, $\Theta = \prod_{t=1}^{N} \Omega_t$ and we define a finite measure μ on Θ, which need not be known by every agent. The marginal measure μ_t of μ on Ω_t is the information distribution of agent t. The measure space $\Lambda_t = (\Omega_t, A_t, \mu_t)$ is called the information space of agent t. The information space is assumed to have the following properties:

- (1) Polish space: every Ω_t is additionally complete;
- (2) it is locally compact: Ω_t is a locally compact space; and
- (3) it is nonatomic: μ_t is a nonatomic measure on A_t.

Let Ω be the set of all possible ω. For each agent t, let $\Phi_t(\omega) : \omega \rightarrow \omega_t$ be an onto mapping defined on Ω . Let $\bar{\omega}_t$ be the signal observed by agent t if ω occurs. Agent t can distinguish between ω' and ω'' if $\Phi_t(\omega') \neq \Phi_t(\omega'')$. If $\Phi_t(\omega) \neq \omega$, the private information space of agent t is called incomplete. Let us define a product of private information space Θ by

$$\Theta = \prod_{t=1}^{N} \Phi_t(\Omega) \tag{1}$$

where $\theta \in \Theta$ is an information set with an explicit representation of the incompleteness of all agents' private information spaces, and call it information structure. The realization of agent t's private information and the total agents' information structure is represented by $\bar{\omega}_t(= \Phi_t(\bar{\omega}))$ and $\bar{\theta} = \prod_{t=1}^{N} \Phi_t(\bar{\omega}) \in \Theta$. Further, we assume that there are some common measures concerning the distribution of information. These measures are represented by a finite measure μ on A:

$$A := \prod_{t=1}^{N} A_t, \ \mu := \bigotimes_{t=1}^{N} \mu_t \tag{2}$$

which need not be known by all individuals. At any moment, let us suppose that these measures are known to all individuals. This assumption will be relaxed in Section 8.4.

8.2.3 Software Space

After observing his information variables each agent t chooses a program from his software space Ξ_t, which is assumed to be Polish and is endowed with its Borel σ-field $B_t := \beta(\Xi_t)$. The measure space $\Upsilon = (\Xi, B)$ is called a software space. The software space is called compact if Ξ is a compact metric space. In order to describe some kind of randomization of strategies, we introduce a mixed strategy for agent t by a stochastic kernel $K_t \in \Gamma_t : \Omega_t \times \Xi_t \rightarrow [0,1]$ by a function $K_t(\cdot,\cdot)$ for which

- (1) $K_t(\omega_t, \cdot)$ is a probability measure on Ξ_t for all $\omega_t \in \Omega_t$; (3)
- (2) $K_t(\cdot, \sigma_t)$ is a measurable function on Ω_t for all $\sigma_t \in \Xi_t$.

For all $t \in T$ a stochastic kernel K_t defines agent t's mixed strategies for his arbitrary private information $\omega_t \in \Omega_t$. Γ_t is a set of stochastic kernels K_t; we call it strategy space. For any stochastic kernel K_t let $\mu_t \otimes K_t$ be the measure on $A_t \otimes B_t$ defined by the measure μ_t and the stochastic kernel K_t. Since both Ω_t and Ξ_t are separable metric spaces, the strategy space Γ_t can be endowed with the corresponding weak topology. A sequence $\{K_t^{(n)} \mid n \in N\}$ from Γ_t is said to converge to $K_t \in \Gamma_t$, if and only if

$$\lim_{n \to \infty} \int h d\mu_t \otimes K_t^{(n)} = \int h d\mu_t \otimes K_t \tag{4}$$

for every bounded continuous function $h : \Omega_t \times \Xi_t \rightarrow R$. The weak topology on Γ_t can be made metric. The space

$$\Gamma := \prod_{t=1}^{N} \Gamma_t \tag{5}$$

is endowed with the corresponding product topology in a metric space, too. Assume that strategy space Γ is a compact metric space.

8.2.4 The Individual's Behaviour

Let us introduce constraint correspondences in that they restrict the set of admissible programs of hardware dependent on information structure $\theta \in \Theta$ and the strategical behaviour of all users $\sigma_{-t} \in \Xi_{-t}$, where $\Xi_{-t} = (\Xi_1, \cdots, \Xi_{t-1}, \Xi_{t+1}, \cdots, \Xi_N)$. The constraint correspondence of agent t is a function $F_t : \Omega_t \times \Gamma \rightarrow C(\Xi_t)$, where $C(\Xi_t)$ is the system of compact non-void subsets of Ξ_t. The behaviour of agent t is assumed to be influenced by the payoff function of agent t, π_t:

$$\pi_t : \Theta \times \Xi \rightarrow R \tag{6}$$

where $\pi_t(\cdot, \sigma)$ is measurable and quasi-concave on Ω_t for given $\sigma_t \in \Xi_t$; i.e., the set $\{\sigma_t \in \Xi_t \mid \pi_t(\theta, (\sigma_t, \bar{\sigma}_{-t})) > \alpha\}$ is convex for every $\alpha \in R, \theta \in \Theta$ and $\bar{\sigma}_{-t} \in \Xi_{-t}$. Define the utility function of agent t, U_t,

$$U_t : R \to R \tag{7}$$

w ere $U_t(\pi_t)$ is a Neuman=Morgenstern utility function and quasi-concave for arbitrary π_t. Suppose every agent $t \in T$ is a utility maximizer and the information of all agents is mutually independent. Assume for the moment $\mu := \bigotimes_{j=1}^{N} \mu_j$ is common knowledge amongst all agents. Then the Jordan-measure $Q(K_{-t} : K_t)$ can be defined by the product of the measures $\mu_t \otimes K_t$ over $A_t \otimes B_t$.

$$Q(K_{-t} : K_t) = \bigotimes_{j=1}^{t-1}(\mu_j \otimes K_j) \otimes (\mu_t \otimes K_t) \otimes \bigotimes_{j=t+1}^{N} (\mu_j \otimes K_j). \tag{8}$$

Then, an agent is assumed to maximize his expected utility $V_t(\bar{K}_{-t} : K_t)$

$$V(\bar{K}_{-t} : K_t^*) = \max_{K_t \in \Gamma_t} \left\{ \int U_t \circ \pi_t(\theta, \sigma) dQ(\bar{K}_{-t} : K_t) \right\} \tag{9}$$

in terms of $K_t \in \Gamma_t$ given others' strategies $\bar{K}_{-t} = (\bar{K}_1, \cdots, \bar{K}_{t-1}, \bar{K}_{t+1}, \cdots, \bar{K}_N)$. With these assumptions, a specified economic environment of a logistical network can be fully described by the 8-tuple of characteristics of the N-tuple of agents; thus a logistical system is generally denoted by

$$\begin{aligned} \Im &= (\Upsilon, \Pi, e) \\ &= \{(\Xi_t, B_t, \pi_t, \Omega_t, A_t, \mu_t, F_t, U_t) : t \in T\}. \end{aligned} \tag{10}$$

The Nash equilibria induced by the situation where all noncooperative agents compete with each other with incomplete information on a network environment fully characterizes our equilibrium concept of a logistical network. Term such equilibrium points 'network equilibria with incomplete information'.

8.3 NETWORK EQUILIBRIA WITH INCOMPLETE INFORMATION

8.3.1 Outline of Previous Studies

Games of this type with differentiated and distributed information were founded by Harsanyi (1967) and have been elaborated by Milgrom and Weber (1985), Radner and Rosenthal (1982) as well as Aumann et al. (1983). The last two contributions are confined to finite decision spaces. Constraint correspondences are not investigated in any of the above contributions. Although games with constraints were originally studied by Wieczorek (1984) in a general set-up, there was no direct connection to games with incomplete information except in the area of equilibrium existence. Meister (1990) has tried to take both constraints and uncertainties into consideration. Aumann et al. and Meister investigated the existence of an equilibrium in randomized strategies and showed some results about the existence of an approximate equilibrium in pure strategies. They also analyzed games which allow exact equilibria in pure strategies. They showed that even in the case of a non-atomic information distribution, such exact equilibria may not exist in pure strategies. However, for certain types of payoff functions and constraints the purification problem proves to have a satisfactory approximate equilibrium. Problems arising in connection with the lack of an information distribution known to all individuals as assumed in the game described hereafter are also discussed in a paper by Mertens and Zamir (1985). In our study, we investigate incomplete information games with Mertens and Zamir types of payoff-functions and generalize the arguments by Mertens to incomplete information games with unknown payoff-functions. This trial can be made by explicitly taking account of the formation of an agent's expectations about his payoff variations. In what follows, we investigate some properties of network equilibria with incomplete information when respective agents have common beliefs about the distribution of the information structure and have perfect information about their payoff-functions. These assumptions are relaxed in subsequent Sections.

8.3.2 Mathematical Preparation

Before investigating the rational expectations equilibria of logistical networks, let us summarize some necessary terminology and useful mathematical arguments. Let $K := (K_1, \cdots, K_N) \in \Gamma$ be the vectors of agents' strategies. A strategy L_t of player t is called an admissible strategy, iff

$$L_t(\omega_t, F_t(\omega_t, K)) = 1 \tag{11}$$

holds. A strategy N-tuple K is called a Nash equilibrium point of the game, iff K_t is an admissible strategy of every agent t, and

$$V_t(K^*_{-t} : K^*_t) \geq V_t(K^*_{-t} : L_t) \quad (t \in T) \tag{12}$$

simultaneously holds for all admissible strategies L_t of all agents. Define the payoff function $\pi : \Theta \times \Xi \rightarrow R^N$ is a μ-Carathéodory function (in short, a μ-C-function); the μ-C-function has the following properties

- (1) $\pi(\cdot,\sigma)$ is A-measurable for all $\sigma \in \Xi$;
- (2) $\pi(\theta,\cdot)$ is continuous for arbitrary $\theta \in \Theta$;
- (3) $|\ \pi_t(\theta,\sigma)\ |\leq Z(\theta)$, for arbitrary $\theta \in \Theta$, $\sigma \in \Xi$, and $t \in T$,

where $Z(\theta)$ is an arbitrary bounded continuous function. Every μ-C-function is $A \otimes B$-measurable and is $\mu \otimes K$-integrable for the arbitrary strategy $K \in \Gamma$.

[Lemma 1] (Meister, 1990): *Assume a μ-C-function $\pi : \Theta \times \Xi \rightarrow R^N$. Then function $V : K \in \Gamma \rightarrow R$ defined by*

$$V(K) = \int U_t \circ \pi d\mu \otimes K$$

is continuous on the set of all strategies $K \in \Gamma$.

We now summarize some important lemmas on μ-C-correspondences. For the system of compact non-void subsets of Ξ, we use the symbol $C(\Xi)$. Define a A-$\beta(C(\Xi))$-measurable correspondence by $F : \Omega \times \Gamma \rightarrow C(\Xi)$.

[Lemma 2] (Meister, 1990): *Let $F : \Omega \times \Gamma \rightarrow C(\Xi)$ be an A-$\beta(C(\Xi))$-measurable correspondence. Then the set*

$$\Gamma_F := \{L \in \Gamma \mid L(\omega, F(\omega, K)) = 1\}$$

is a nonempty convex and compact subset of Γ.

[Lemma 3] (Meister, 1990): *Assume a metric space, Γ. Let $F : \Omega \times \Gamma \rightarrow C(\Xi)$ be a correspondence with the properties*

(1)$F(\theta,\cdot)$ is continuous at arbitrary $K \in \Gamma$ for all $\theta \in \Theta$;

(2)$F(\cdot,K)$ is A-$\beta(C(\Xi))$-measurable at arbitrary $K \in \Gamma$.

Then, the correspondence $\xi(K) := \Gamma_{F(\cdot,K)}$ is continuous at $K \in \Gamma$.

8.3.3 Existence Theorem of Incomplete Information Equilibria

From the above Lemmas, we directly obtain the next Theorem.

[Theorem 1]:*Suppose that the logistical system L has the following properties:*

(1) (Ω_t, A_t, μ_t) is Polish and a locally compact information space;

(2) μ_t *is a probability measure;*

(3) $\Xi_t : t \in T$ *is a compact metric space;*

(4) $F_t : \Omega_t \times \Gamma \to C(\Xi_t)$ *is a Carathéodory constraint correspondence; i.e.,* $F_t(\cdot, K)$ *is an* A-$\beta(C(\Xi_t))$*-measurable correspondence for every* $K \in \Gamma$, *and* $F_t(\omega_t, \cdot)$ *is a continuous correspondence for every* $\omega_t \in \Omega_t$ *for every agent* t;

(5) $\pi_t(\theta, \sigma)$ *is quasi-concave in terms of* σ;

(6) U_t *is quasi-concave.*

Then, the logistical network has an equilibrium point.

Proof: Lemma 2 shows the space $\Gamma := \prod_{t=1}^{N} \Gamma_t$ is nonempty, convex and compact. Further, setting

$$\xi_t(K) : = \{L_t \in \Gamma_t \mid L_t(\omega_t, F_t(\omega_t, K)) = 1\} \tag{13}$$

for $K \in \Gamma$, as a consequence of Lemmas 1 and 3, we obtain a continuous, convex- and compact-valued correspondence $\xi_t : \Gamma_t \to C(\Gamma_t)$. Let $\xi^* : \Gamma \to C(\Gamma)$ be given by

$$\xi^*(K) := \prod_{t=1}^{N} \{L_t \in \xi_t(K) \mid V_t(K_{-t} : L_t) = \max_{L_t'} V_t(K_{-t} : L_t')\}.$$

Then, because V_t is continuous from Lemma 1, ξ^* is a non-empty compact-valued, upper semi-continuous correspondence. Moreover, since V_t is quasi-concave in terms of L_t, ξ^* is convex-valued. Then, Ky Fan's Fixed Point Theorem (Fan, 1952) applies to ξ^* and delivers a strategy $K^* \in \Gamma$ with $K^* \in \xi^*(K^*)$. By definition of ξ_t and ξ^*, $K \in \Gamma$ is admissible for every agent $t \in T$ and satisfies

$$V_t(K_{-t}^* : L_t) \leq V_t(K_{-t}^* : K_t^*) \tag{14}$$

for arbitrary admissible strategies $L_t \in F_t(\omega_t, K)$. Hence, K^* is an equilibrium point. (Q.E.D.)

8.4 RATIONAL EXPECTATIONS EQUILIBRIA

8.4.1 Rational Expectations Hypothesis

In our arguments about incomplete information games, we assumed that every agent possesses a common belief about the distribution of information structure $\mu = \prod_{t=1}^{N} \mu_t$ and has perfect information about their payoff-functions. However, in a game with many agents, it is very unlikely that agents would have such rich information to calculate their payoff-functions. Rather, it is natural to assume that in the situation where agents repeat their choices of programs with different information, they will have opportunities to learn about others' actions, since current payoffs might reflect - in a possibly complicated manner - the private information signals received by other agents and other agents' actions. Thus, a rational agent could form expectations by acquiring and processing information before he makes his decisions.

The past decade has witnessed important developments in the study of the expectations formation process and the problem of decision-making under uncertainty. Of the theories of expectations formation so far advanced, the rational expectations hypothesis has attracted by far the greatest attention. The rational expectations hypothesis (REH) due to Muth (1961) states that subjective expectations held by economic agents will be the same as conditional mathematical expectations based on the true probability model of the economy; or more generally - that the agents' subjective probability distribution coincides with the objective probability distribution of events.

Although the REH was advanced by Muth it was work of the Lucas (1978), Sargent (1973), Barro (1976) and others that brought it into prominence. In the REH it assumed that an agent who has a good understanding of logistical networks is in a position to use the distribution of his payoff to make inferences about the actions taken by other agents. These inferences are derived, explicitly or implicitly, from an individual's model of the relationship between the private information received by himself and the payoff-values. On the other hand, the true relationship is determined by the individual agents' behaviour, and hence by their individual models. Agents have the opportunity to revise their individual models in the light of observations. Hence, there is a feedback from the true relationship to the individual models. An equilibrium of this system, in which the individual models are identical to the true model, is called a rational expectations equilibrium (REE). In what follows, we investigate some properties of network equilibria when respective agents may form rational expectations about their payoff-value distribution.

8.4.2 Equilibria with Incomplete Information about Payoff Functions

As we stated in the previous Section, Nash equilibria for incomplete information games (Bayesian equilibria) may involve a situation in which an agent randomizes his acts; that is, the equilibrium may call for an agent, in some circumstances, to choose a program according to some probability distribution. To answer this criticism, it has been suggested that the concept of the purification of mixed-strategy equilibria be employed. Aumann et al (1983) showed that if individual agents have some private information, then pure-strategy equilibria will exist. An example of pure-starategy equilibria will be discussed in Section 6.

If an agent sees the outcome of a random variable that is rich enough and independent of everything other agents see, his act can be a function of his private information. In what follows, we investigate some properties of incomplete information games with unknown payoff functions and show the existence of the rational expectations equilibria. Suppose that the preferences of agents are derived from the payoff functions:

$$\pi_t : \prod_{j=1}^{N} \Omega_j \times \prod_{j=1}^{N} \Xi_j \rightarrow R \quad (t \in T). \tag{15}$$

Since agent t does not know exactly his counter agents' strategy space, the game can also be considered as a game with unknown payoff functions. Let us assume an information distribution measure μ on $A = \bigotimes_{j=1}^{N} A_j$, which we assume to be known to all agents (this assumption is relaxed later). Further assume that μ is a finite measure with a density function $f : \prod_{t=1}^{N} \Omega_t \rightarrow R_+$ by which the Jordan measure on the product measures $\mu_1 \otimes \cdots \otimes \mu_N$ can be represented by

$$\mu = f(\mu_1 \otimes \cdots \otimes \mu_N) < \infty. \tag{16}$$

We claim that π_t is a μ-C-function for every agent t. For every mixed strategy $N-$ tuple $(K_1, \cdots, K_N)$ let the stochastic kernel $K_1 \otimes \cdots \otimes K_N$ be defined by

$$\begin{aligned} K_1 \otimes \cdots \otimes K_N (\omega_1, \cdots, \omega_N, \sigma_1, \cdots, \sigma_N) \\ := K_1(\omega_1, \sigma_1) \times \cdots \times K_N(\omega_N, \sigma_N), \\ \theta = (\omega_1, \cdots, \omega_N) \in \Theta, \ \sigma_t \in \Xi_t. \end{aligned} \tag{17}$$

The function $V_t : \Gamma \rightarrow R$ defined by

$$V_t(K_1, \cdots, K_N) := \int U_t \circ \pi_t(\theta, \sigma) d\mu \otimes (K_1 \otimes \cdots \otimes K_N) \tag{18}$$

is called the expected utility function of agent t.

Let us now define the incomplete information equilibria of the logistical network $\Im$. A strategy $N-$tuple $K^* := (K_1^*, \cdots, K_N^*)$ is called an equilibrium point of $\Im$, if K_t is admissible w.r.t. K^*, and

$$\begin{aligned} V_t(K_1^*, \ \cdots, K_{t-1}^*, L_t, K_{t+1}^*, \cdots, K_N^*) \\ \leq V_t(K_1^*, \cdots, K_{t-1}^*, K_t^*, K_{t+1}^*, \cdots, K_N^*) \end{aligned} \tag{19}$$

holds for every admissible strategy $L_t \in F_t(\omega_t, K)$ and every agent t. From Theorem 1, we get the next Corollary.

[Corollary 1]: *Assume that a logistical network L holds the conditions of Theorem 1 and has the properties identified by Equations (16) and (17). Then, there exists an equilibrium point in L.* (For the proof, see Meister, 1990).

Let us characterize the equilibrium point of L with unknown payoff functions. Introduce the conditional payoff function $W_t(\omega_t, \sigma_t; K_{-t}) : \Omega_t \times \Xi_t \rightarrow R$ of agent t by

$$W_t(\omega_t, \sigma_t; K_{-t}) := \int U_t \circ \pi_t(\theta, \sigma) f(\theta) d \bigotimes_{j \neq t} (\mu_j \otimes K_j) \tag{20}$$

where $\theta = (\omega_1, \cdots, \omega_N) \in \prod_{j=1}^N \Omega_j$, $K_{-t} = (K_1, \cdots, K_{t-1}, K_{t+1}, \cdots, K_N) \in \prod_{j \neq t} \Gamma_j$ and $\sigma = (\sigma_1, \cdots, \sigma_N) \in \prod_{j=1}^N \Xi_j$. Thus, we get

$$V_t(K_{-t} : K_t) = \int W_t(\omega_t, \sigma_t; K_{-t}) d\mu_t \otimes K_t. \tag{21}$$

As a consequence of (16) and (17), the conditional payoff $W_t(\omega_t, \sigma_t; K_{-t})$ is a μ_t-C-function. Thus, for the unknown payoff functions, we also have the following Corollary.

[Corollary 2]: *A strategy N-tuple $K^* = (K_1^*, \cdots, K_N^*)$ is an equilibrium point, if and only if K_t^* is K-admissible for every agent t and*

$$V_t(K_{-t}^* : K_t^*) = \max_{K_t \in F_t(\omega_t, K^*)} V_t(K_{-t}^* : K_t) \tag{22}$$

hold for $\omega_t \in \Omega_t$-a.e and $K_t \in F_t(\omega_t, K^*)$.

8.4.3 Existence of REE

In the arguments of the previous Section, every agent is assumed to possess a common belief about the distribution of the information structure. In what follows, this assumption will be relaxed. Every agent is assumed to be in a position to use the distribution of his payoffs to make inferences about the actions taken by other agents. These inferences are derived, explicitly or implicitly, from an individual's model of the relationship between the private information received by himself and the payoff-values.

From these assumptions, payoff functions $\pi : \Theta \times \Xi \rightarrow R^N$ are μ-C-functions, i.e. they are $A \otimes B$-measurable, $\mu \otimes K$-integrable for arbitrary $K \in \Gamma$, and continuous for every $\theta \in \Theta$. Therefore, naturally the Jordan measure on the image space of the payoff function can be defined. Let (Δ, ρ) be a measurable space on the payoff space R^N, and

ρ is a Borel σ-field $\rho = \beta(\Delta)$. Since payoff function π is a measurable function from the measure space $((\Theta \times \Xi), (A \otimes B), (\mu \otimes K))$ onto measurable space (Δ, ρ), we have $\pi^{-1}(\rho) \subset (A \otimes B)$. It is obvious that measure space $((\Theta \times \Xi), (A \otimes B), (\mu \otimes K))$ constitutes a probability space, i.e., $\int_{\Theta \times \Xi} d\mu \otimes K = 1$. Let the image function $\tau(v; K)$ be

$$\tau(v; K) = (\mu \otimes K) \circ \pi^{-1} : \Delta \to R^N \tag{23}$$

for arbitrary K-admissible stochastic kernels $K \in F(\omega, K(\omega))$ and $v \in \rho$. Then, $\tau(v; K)$ defines image measures (probability measures) on measurable space (Δ, ρ). The measure space (Δ, ρ, τ) is called the probability space induced from the probability space $((\Theta \times \Xi), (A \otimes B), (\mu \otimes K))$ by a measurable function π. Note that $\{\pi^{-1}(v) : v \in \rho\}$ is a class of the subsets of $(A \otimes B)$ and is σ-additive. Call it the σ-additive class induced by π and denote it by $(A \otimes B)_\pi$.

[Lemma 4] (Lehmann, 1959): *Let $((\Theta \times \Xi), (A \otimes B))$ and (Δ, ρ) be a measurable space and $\pi : (\Theta \otimes \Xi) \to \Delta$ be measurable, i.e. $\pi^{-1}(v) \subset (A \otimes B)$. Then the function $U^* : (\Theta \times \Xi) \to R$ is $(A \otimes B)_\pi$-measurable relative to the σ-algebra π^{-1} iff there is a measurable function $U : \Delta \to R$ such that $U^* = U \circ \pi(\theta \times \sigma)$ for every $(\theta \times \sigma) \in (\Theta \times \Xi)$.*

From Lemma 4, we directly have the next Lemma.

[Lemma 5]: *Let $((\Theta \times \Xi), (A \otimes B), (\mu \otimes K))$ be a measure space and (Δ, ρ) be measurable space and $U_t : \Delta \to R$ be ρ-measurable. If $\tau := (\mu \otimes K) \circ \pi_t^{-1} : \Delta \to R_+$ is the image function, then for each $v \in \rho$ and arbitrary $K \in F(\omega, K)$, we have*

$$\begin{aligned}\int_{(\Theta \times \Xi)} & U_t \circ \pi_t(\omega, \sigma) d\mu \otimes (K_1 \otimes \cdots \otimes K_N) \\ &= \int_\Delta U_t(v_t) d\tau_t(v_t; K).\end{aligned} \tag{24}$$

Proof: The Lemma is easily deduced from the definition of the image measure. Indeed if $U_t(v_t) = \lambda_{A_{it}}(v_t)$, then the left side of (24) becomes

$$\int_{(\Theta \times \Xi)} \lambda_{A_{it}} \circ \pi_t(\omega, \sigma) d\mu \otimes (K_1 \otimes \cdots \otimes K_N)$$

$$= \int_{(\Theta \times \Xi)} \lambda_{\pi_t^{-1}(A)} d\mu \otimes (K_1 \otimes \cdots \otimes K_N)$$

$$= \int_\Delta \lambda_{A_{it}} d\tau_t(v_t; K) = \int_\Delta U_t(v_t) d\tau_t(v_t; K).$$

Thus (24) is true and by the linearity of the integral and the σ-additivity of τ_t the same result holds if $U_{tn} = \sum_{i=1}^{n} \alpha_i \lambda_{A_{it}}$, a step function with $\alpha_i \geq 0$. If $U_t \geq 0$ is measurable, then there exist step functions $0 \leq \lim_{n\to\infty} U_{tn} \to U_t$, so that (24) holds by the Lebesgue monotone convergence theorem. Since any measurable $U_t = U_t^+ - U_t^-$ with $U_t^+, U_t^- \geq 0$ and is measurable, the last statement implies the truth of (24) in general. (Q.E.D.)

Assume $\bar{K}_{-t} \in \Xi_{-t}$. Define the conditional image measure for agent t by $\tau_t(v_t; K_t, \bar{K}_{-t}) : \Delta \to R_+$

$$\tau_t(v_t; K_t, \bar{K}_{-t}) := Q(\bar{K}_{-t} : K_t) \circ \pi^{-1} \tag{25}$$

for arbitrary $K_t \in F_t(\omega_t, K)$, where $Q(\bar{K}_{-t} : K_t) = \mu \otimes (\bar{K}_1 \otimes \cdots \otimes \bar{K}_{t-1} \otimes K_t \otimes \bar{K}_{t+1} \otimes \cdots \otimes \bar{K}_N)$. Then, from Corollary 2, Lemmas 4 and 5 we have the next Theorem.

[Theorem 2]: *Let the image function $\tau_t(v_t; K_t, \bar{K}_{-t}) : \Delta \to R_+$ be ρ-measurable and let the utility function $U_t : \Delta \to R$ be $(A \otimes B)_\pi$-measurable. Then, there exists an equilibrium $K^* = (K_1^*, \cdots, K_N^*) \in F(\omega, K^*)$, which holds for all $t \in T$ and arbitrary $L_t \in F_t(\omega_t, K^*)$*

$$(1) \int_{(\Theta\times\Xi)} U_t \circ \pi_t(\omega, \sigma) dQ(\bar{K}_{-t}^* : K_t^*) = \int_\Delta U_t(v_t) d\tau_t(v_t; K_t^*, \bar{K}_{-t}^*) \tag{26}$$

$$(2) \int_\Delta U_t(v_t) d\tau_t(v_t; K_t^*, \bar{K}_{-t}^*) \geq \int_\Delta U_t(v_t) d\tau_t(v_t; L_t, \bar{K}_{-t}^*). \tag{27}$$

Let the function $\tau_t^e(v_t; L_t) : \Delta \to R_+$ represent the expectations of agent t on the conditional image measures. Assume that each agent learns the variation of his payoffs through the recurrence of his choice, and eventually his expectations $\tau_t^e(v_t; L_t)$ coincide with the objective image measures $\tau_t(v_t; K_t^*, \bar{K}_{-t}^*)$ in the long run; he has formed the rational expectations on the variation of his payoffs. Thus, the rational expectations of agent t can be described by

$$\tau_t^*(v_t; K_t^*) := Q(\bar{K}_{-t}^* : K_t^*) \circ \pi_t^{-1} \tag{28}$$

which coincide with the image measure which will be realized in the long run. Theorem 2 manifests that there exist rational expectations equilibria (REE) where every agent's subjective expectations for his payoff variation coincide with the objective distribution of payoff realized in the logistical network in the long run.

8.5 REE IN DISCRETE LOGISTICAL NETWORKS

Consider the logistical networks with the discrete hardware space S whose density is finite. The software space is the set of programs which is characterized by the class of subsets of S. Thus the software space can be represented by discrete measurable space $\Xi_0 = (C(S), S)$. Define the individual software space of agent t by $\Xi_{0t} = (C(S_t), S_t)$. Let $K_0 \in \Gamma_0$ be discrete strategies and Γ_0 be a discrete strategy space. A strategy $K_0 \in \Gamma_0$ is called discrete iff K is of the type

$$K_{0t}(\omega_t, \sigma_{0t}) = \sum_{i=1}^{r} \alpha_{it}(\omega_t)\varepsilon_{f_{it}}(\omega_t) \tag{29}$$

for some $r \in N$ (the density of software space) and some measurable function $f_{it}(i = 1, \cdots, N) : \Omega_t \to \Xi_{0t}$ and the probability function $\alpha_{it}(\omega_t)(i = 1, \cdots, r) : \Omega_t \to [0,1]$ with $\sum_{i=1}^{N} \alpha_{it}(\omega_t) = 1$. For a strategy $K_{0t} \in \Gamma_{0t}$ of this type we introduce the symbol

$$K_{0t} : \gamma_t(f_1, \cdots, f_r, \alpha_1, \cdots, \alpha_r)(\omega_t, \sigma_{0t}).$$

The expected utility function of agent t is given by

$$V_t(\gamma_t) = \int U_t \circ \pi_t(\theta, \sigma_0) \prod_{j=1}^{\infty} d\mu_j \otimes \gamma_j(\omega_j, d\sigma_{0j}). \tag{30}$$

Then, if the strategies $\gamma \in \Gamma_0$ satisfy

$$V_t(\gamma_1^*, \cdots, \gamma_{t-1}^*, \gamma_t', \gamma_{t+1}^*, \cdots, \gamma_N^*) \leq V_t(\gamma_1^*, \cdots, \gamma_N^*) \tag{31}$$

for arbitrary $\gamma_t' \in \Gamma_{0t}$ of every $t \in T$, $\gamma \in \Gamma_0$ is called the discrete equilibria.

[Lemma 6] (Meister,1990): *For arbitrary $K_0 \in \Gamma_0, \pi \in \Re^N$ and a correspondence with a measurable graph $F : \Omega \to \Xi_0$, assume that $K_{0t}(\omega, F(\omega)) = 1$ holds. Then, there exist discrete strategies $\gamma(f_1, \cdots, f_r, \alpha_1, \cdots, \alpha_r)$ which satisfy*

$$(1) \int U_t \circ \pi_t(\theta, \sigma_0) d\mu \otimes \gamma(\theta, d\sigma_0)$$
$$= \int U_t \circ \pi_t(\theta, \sigma_0) d\mu \otimes K_0(\theta, d\sigma_0)$$

$$(2)\ f_1(\theta), \cdots, f_r(\theta) \in F(\theta)$$

for arbitrary $\theta \in \Theta$. $\Re^N$ is a set of arbitrary measurable functions $\pi : \Theta \times \Xi_0 \to R^N$.

From Lemma 6, we understand that there exists a discrete equilibrium in our logistical network. The equilibrium problem of logistical networks can be regarded as a special form of the problems discussed in the previous Section. Thus, from Theorem 2, we can directly derive the next Corollary.

[Corollary 3]: *Suppose the same conditions as in Theorem 2. Then, there exists an equilibrium point* $\{\gamma_t^* \in F_t(\omega_t, \gamma) : t \in T\}$, *which holds for all* $t \in T$ *and arbitrary* $\gamma_t' \in F_t(\omega_t, \gamma)$

$$(1) \int_{(\Theta\times\Xi_0)} U_t \circ \pi_t(\omega, \sigma) dQ_0(\bar{\gamma}_{-t}^* : \gamma_t^*) = \int_\Delta U_t(v_t) d\tau_t(v_t; \gamma_t^*, \bar{\gamma}_{-t}^*) \quad (32)$$

$$(2) \int_\Delta U_t(v_t) d\tau_t(v_t; \gamma_t^*, \bar{\gamma}_{-t}^*) \geq \int_\Delta U_t(v_t) d\tau_t(v_t; \gamma_t, \bar{\gamma}_{-t}^*) \quad (33)$$

where $Q_0(\bar{\gamma}_{-t}^* : \gamma_t^*) = \mu \otimes (\bar{\gamma}_1^* \otimes \cdots \otimes \bar{\gamma}_{t-1}^* \otimes \gamma_t^* \otimes \bar{\gamma}_{t+1}^* \otimes \cdots \otimes \bar{\gamma}_N^*)$.

8.6 EXAMPLE OF REE

8.6.1 Specification of a Network Structure

Consider a discrete logistical network with a finite number of components. Denote the set of admissible programs by Ξ and the set of agents by $T = \{1, \cdots, N\}$. The set of hardware components available for a program a is denoted by κ_a. Every agent is assumed to adopt a pure strategy. Consider the situation where one agent t chooses a program $a \in \Xi_t$ with his private information, $\bar{\omega}_t$, and others choose pure strategies $\bar{\gamma}_{-t}$. Denote the N-tuple of pure strategies in the above setting by $\bar{\gamma}_a(\theta) = \{\bar{\gamma}_1(\bar{\omega}_1), \cdots, \gamma_{t-1}(\bar{\omega}_{t-1}), a, \gamma(\bar{\omega}_{t+1}), \cdots, \gamma(\bar{\omega}_N)\}$. Let us specify the expected utility function of agent t for program a by

$$V_{at}(\bar{\omega}_t) = \int U(\pi_a(\bar{\gamma}_a(\theta))) d\mu + \bar{\omega}_{at} \quad (34)$$

where $d\mu := \prod_{t=1}^N d\omega_t$ and π_a is the profit driven by choosing program a. Moreover, assume $E[\omega_{at} \cdot \omega_{a't'}] = 0$ $(a_t \neq a_{t'} \in \Xi_{0t}; t, t' \in T)$. If private information is mutually independent, no private information conveys any information about others' behaviour. Denote the private information related to program a by a probability variable ω_{at} and assume it varies according to the I.I.D. Weibull distribution with the density function $f(\omega_{at}) = \lambda \exp(-\lambda\omega_{at}\exp(-\exp(-\lambda\omega_{at})))$.

On the other hand, the rational expectations about the distribution of agent t's payoff can be given by the normal distribution, χ^*, with mean μ_a^* and variance σ_a^{*2}. Later, we will show RE χ^* is given by them. Suppose a utility function with the constancy of absolute risk aversion. Then the utility function takes the form, $U(\tau) = -\exp(-\zeta\tau)$. The expected utility for program a under χ^* and private information $\bar{\omega}_{at}$ is approximated by

$$V_{at}(\bar{\omega}_{at} : \chi^*) = \mu_a^* + \eta(\sigma_a^{*2}) + \bar{\omega}_{at}, \tag{35}$$

where $\eta(\sigma_a^{*2}) = -\sigma_a^{*2}U''/2U'$ (> 0): the risk premium of agents. Under the constancy of the absolute risk aversion measure, it holds that $\eta(\sigma_a^{*2}) = -\zeta\sigma_a^{*2}$ (*constant*). The program chosen by agent t with REχ^* and private information $\bar{\omega}_t$ is given by

$$\gamma_{at}^*(\bar{\omega}_{at} : \chi^*) = arg \max_a \{V_{at}(\bar{\omega}_{at} : \chi^*)\}. \tag{36}$$

The terms μ_a^*, σ_a^{*2} included in the R.H.S. of Equation (35) are unknown variables, whose values are endogenously determined to coincide with the objective values realized in the network in the long run. From Corollary 3 and Equation (34), there should exist χ^* which satisfies the conditions:

$$\gamma_t^*(\bar{\omega}_{at} : \chi^*) = arg \max_a \left\{ \int U[\pi_a(X(\theta : \chi^*))]\psi(\theta)d\mu + \bar{\omega}_{at} \right\} \tag{37}$$

for arbitrary $t \in T$ and $\omega_t \in \Omega_t$, where $\pi_a(X)$: the payoff function for program a, $X(\theta : \chi^*) := \{\sum_{t\in T} \delta_a(\gamma_t^*(\omega_t : \chi^*)) : a \in \gamma_0\}$: the number of users of programs (vector) and δ_a: $0-1$ indices, where $\delta_a = 1$, if program $\gamma_t^*(\omega_t : \chi^*)$ coincides with program a; $\delta_a = 0$, if otherwise. In other words, the equilibrium point can be calculated by finding μ_a^*, σ_a^{*2} which simultaneously satisfy Equation (37) for arbitrary $t \in T$ and $\omega_t \in \Omega_t$.

8.6.2 REE Network Equilibrium Model

Let N be the number of users of the network and M be the number of programs. Assume that the set of programs available for every agent is identical, and denote it by Ξ. The number of users of respective programs is given by M-triple vector $q = (q_1, \cdots, q_M)$. The private information ω_{at} is independently and identically distributed according to the Weibull Distribution. Then the probability for each agent to choose program $a \in \Xi$ is given by

$$\begin{aligned} P_a &= \text{Prob}\{EU_a \geq EU_b; b \in \Xi\} \\ &= \frac{\exp(\lambda EU_a)}{\sum_{b\in\Xi}\exp(\lambda EU_b)} \end{aligned} \tag{38}$$

where $EU_a = \mu_a - \zeta\sigma_a^2/2$. The distribution of the number of users of respective programs can be derived. The probability of obtaining a given choice pattern of programs, $q = (q_1, \cdots, q_M)$, is given by the multinomial density function with $P = (p_1, \cdots, p_M)$. In other words

$$P(q) = N! \prod_{a=1}^{M} (p_a)^{q_a}/q_a! \tag{39}$$

where $\sum_{a=1}^{M} q_a = N$. The moments of the distribution of q are

$$\begin{aligned} E[q_a] &= NP_a(1-P_a) \\ \text{VAR}[q_a] &= NP_a(1-P_a) \\ \text{COV}[q_a, q_{a'}] &= -NP_aP_{a'} \qquad (a \neq a'). \end{aligned} \tag{40}$$

In principle, the joint density function of the choice patterns can be approximated from (40) by a multivariate normal distribution. To do this, the joint distribution of all choice pattern is first approximated by a single multivariate normal distribution function, $MVN(q^*, \Sigma^*)$, where $q^* = (E[q_1], \cdots, E[q_M])$, and Σ^* is a block diagonal covariance matrix whose elements are given by (40).

Assume that each hardware component can bear its own user benefits (costs). The payoff function of the hardware component z which represents the benefits driven by the usage of the components z is given by the linear benefit function:

$$\pi_z = \alpha_z + \beta_z \left(\sum_{a\in\Xi} \delta_{a,z} q_a \right) \tag{41}$$

where $\delta_{a,z}$: 0-1 indices representing that $\delta_{a,z} = 1$, if program a utilizes component z; $\delta_{a,z} = 0$, if otherwise. The objective variation of the payoff-values can be described by a multivariate-normal distribution with mean $\bar{\mu}_a$ and variance $\bar{\sigma}_z^2$:

$$\begin{aligned} \bar{\mu}_z &= \alpha_z + \beta_z N \sum_{a\in\Xi} \delta_{a,z} P_a \\ \bar{\sigma}_z^2 &= \beta_z^2 N \sum_{a\in\Xi} \delta_{a,z} P_a (1-P_a). \end{aligned} \tag{42}$$

Then, the objective payoff-distribution of program $a \in \Xi$ is characterized by the multivariate normal distribution with parameters:

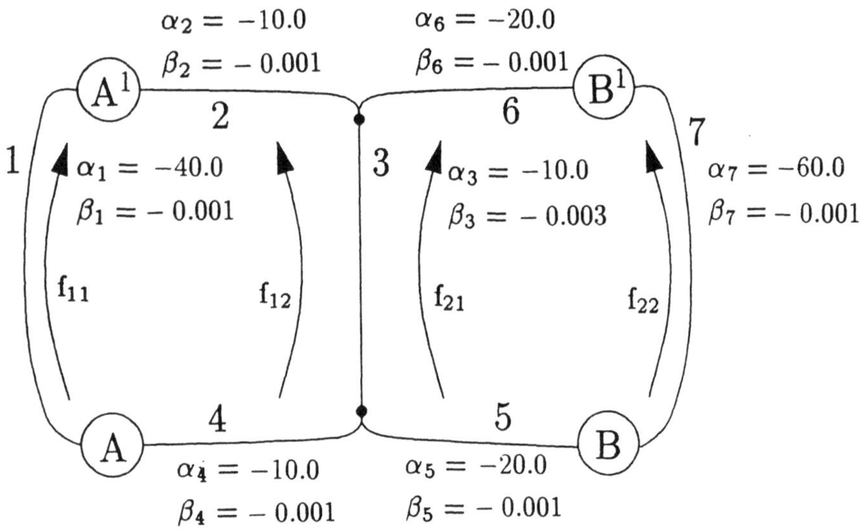

Figure 8.1 : Hypothetical network with two O-D pairs

$$\bar{\mu}_a = \sum_{z \in \kappa_a} \bar{\mu}_z \ , \ \bar{\sigma}_a^2 = \sum_{z \in \kappa_a} \bar{\sigma}_z^2. \tag{43}$$

Show explicitly that parameters $\bar{\mu}_a$ and $\bar{\sigma}_a^2$ are functions of the choice probability vector P, and rewrite them as $\bar{\mu}_a(P)$ and $\bar{\sigma}_a(P)^2$. From Collorary 3, the rational expectations equilibria can be given by the solutions $(\bar{\mu}_a(P^*), \bar{\sigma}_a^2(P^*))$ which, for every $a \in \Xi$, simultaneously satisfy the equations

$$\gamma_t^*(\bar{\omega}_t : \chi^*) = arg \max_a \{EU_a(P^*) + \bar{\omega}_t\} \tag{44}$$

where $EU_a(P^*) = \bar{\mu}_a(P^*) - \zeta \bar{\sigma}_a(P^*)^2/2$. Since Equation (44) holds for every $\omega_t \in \Omega_t$, integrate both sides of Equation (44) in terms of $\bar{\omega}_t$. Then, REE can be given by $P^* = \{P_a^* : a \in \Xi\}$ satisfying simultaneously

$$P_a^* = \frac{\exp\{\lambda EU_a(P^*)\}}{\sum_{b \in \Xi} \exp\{\lambda EU_b(P^*)\}} \tag{45}$$

for all $a \in \Xi$. Equation (45) shows that the necessary conditions for the agents' subjective expectations about the payoff distribution of programs coincides with its objective distribution for every program at the equilibrium point. Note that our equilibrium concepts only

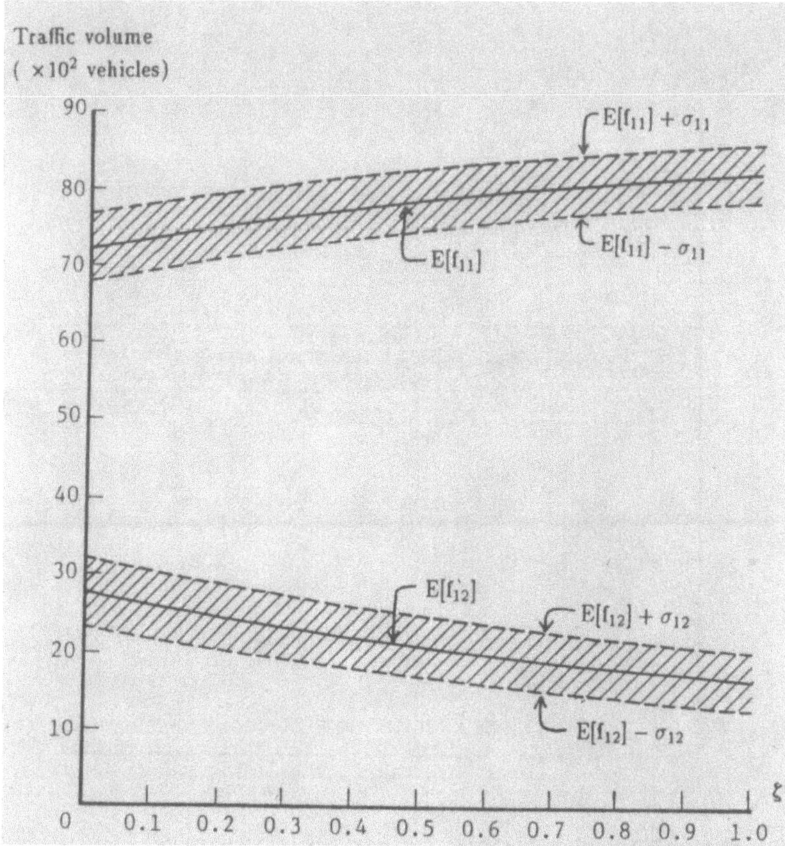

Figure 8.2 : The equilibrium share of flows on respective routes versus the values of ARA-measures for Case 1 ($\lambda = 0.3$)

regulate that there should exist an ex ante equilibrium in agents' subjective beliefs about payoff distribution. The acts chosen are the best responses for each other for every $\omega_t \in \Omega_t$. The existence of such an ex post equilibrium is, of course, not guaranteed by the existence theorem (Postlewaite and Schmeidler, 1987).

8.6.3 Numerical Examples

Consider the simple discrete (traffic) network shown in Figure 8.1. In this example, each route connecting an O-D pair corresponds to the program and the set of routes forms a software space (strategy space). This network includes seven links and four routes connecting two O-D pairs. The agents with different O-D pairs may choose different sets

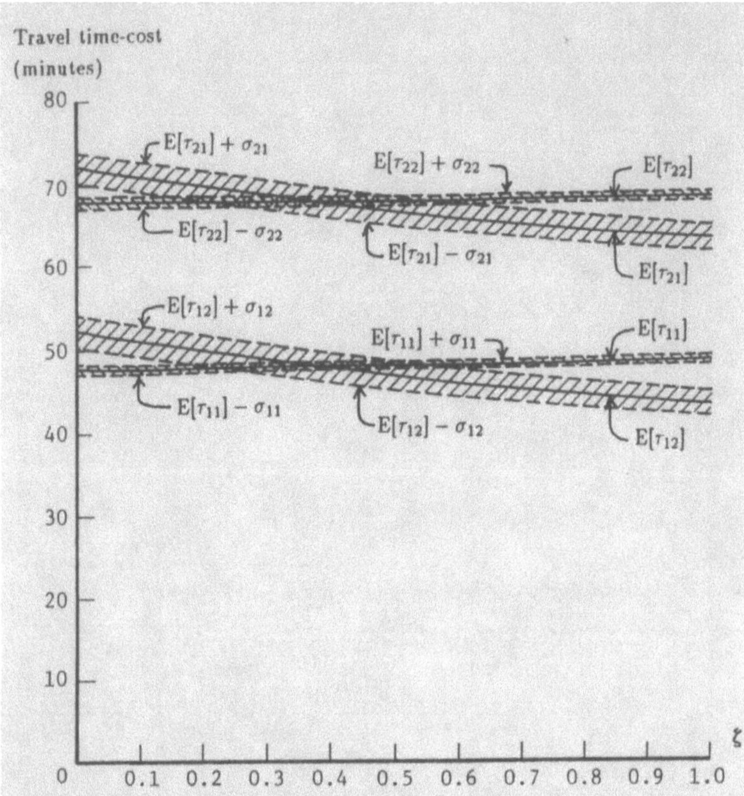

Figure 8.3 : The travel time-costs of respective routes versus the values of ARA-measures for Case 1 ($\lambda = 0.3$)

of alternative programs. Thus, traffic networks can be regarded as typical examples of discrete logistical networks.

Let us investigate the properties of the REE defined by Equation (45) by numerical experiments. Denote the R.H.S. of Equation (45) by $F(P) = \{f_a(P) : a \in \Xi\}$. $F(P)$ is a continuous mapping defined on the compact set $D = \{P \mid \sum_a P_a = 1\}$. From Brouwer's Fixed Point Theorem, there exists an equilibrium in Equation (45). The rink performance functions are given by linear functions: $v_z = \alpha_z + \beta_z(\sum_{a \in \Xi} \delta_{a,z} q_a)$, whose parameter values are summarized in Figure 8.1. Set the two calculation cases: 1) Agents have homogenous preferences (Case 1), 2) Agents have heterogeneous preferences (Case 2). In Case 2, agents are classified into two groups based on their attitudes towards risks, i.e., a) Group A: those who have risk neutral preferences and b) Group B: those who are risk aversive.

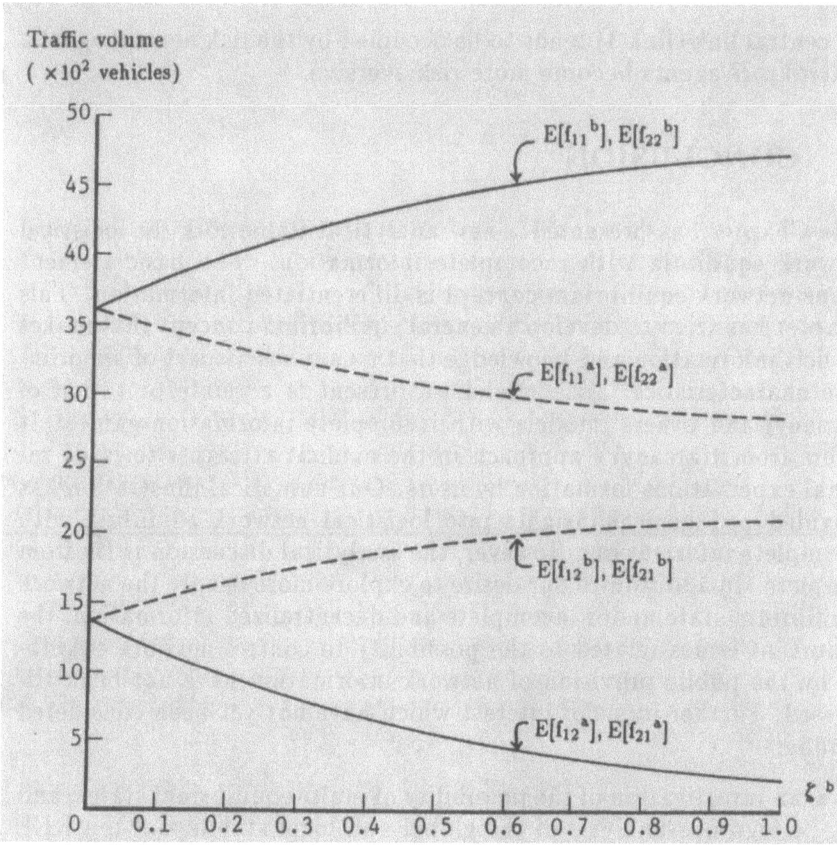

Figure 8.4 : The equilibrium share of flows on respective routes versus the value of ARA-measures for Case 2 ($\lambda = 0.3$)

(1) Homogeneous Preferences (Case 1)

Let the number of agents of O-D pairs (A-A') and (B-B') be 10,000 units, respectively. Define four routes as shown in Figure 8.1 and denote the number of users of each route by $f = (f_{11}, f_{12}, f_{21}, f_{22})$. Figure 8.2 explains the relationships between the route volume f and the ARA (Absolute Risk Aversive)-measures, $\zeta = -U''/U'$. As the values of ζ become larger, fewer agents tend to choose the routes including link 3, which has a large β_z. Figure 8.3 shows how the travel time-costs of routes change with the agents' ARA-measures.

(2) Heterogeneous Preferences (Case 2)

Figure 8.4 describes the relationships between the number of users of routes and the ARA-measures ζ_b of agents in Group B. If agents in Group B are more risk aversive, i.e., ζ_b-values become larger, they become unwilling to utilize the risky routes; f_{12}^b, f_{21}^b decrease. Thus, the central link (link 3) tends to be occupied by the risk neutral agents, as Group B agents become more risk aversive.

8.7 CONCLUSION

This Chapter has presented a new analytical framework for logistical network equilibria with incomplete information. The basic element of our network equilibrium concept is differentiated information. This Chapter has tried to develop a general equilibrium concept that makes explicit information and knowledge that a user has as part of his primitive characteristics. The model we present is a reinterpretation of Harsanyi and others' models with incomplete information games. It differs from Harsanyi's approach in the explicit attempts towards rational expectations formation by users. Our numerical illustration has provided pedagogical insights into logistical network equilibria with incomplete information. However, the analytical discussion is far from complete. In addition to our desire to explore more deeply the network equilibrium state under incomplete and decentralized information, the important issues related to the possibility to control network equilibria by the public provision of network information have not been addressed. Further items of interest which have not yet been considered include:

- an investigation of the possibility of multi-equilibrium states, and a rigorous analysis of the global and local stability of the REE of logistical networks;
- the development of an analytical framework to describe the agents' process of rational expectations formation;
- an investigation of the informational role of common information from public or private sources.

REFERENCES

Andersson, Å.E., 1986, "Presidential address; The four logistical revolutions", *Papers of the Regional Science Association,* 59:1-12.

Aumann, R.J., 1976, "Agreeing to disagree", *Annals of Statistics,* 4:1236-1239.

Aumann, R.J. et al., 1983, "Approximate purification of mixed strategies", *Mathematics of Operations Research,* 8:3:327-341.

Barro, R.J., 1976, "Rational expectations and the role of monetary policy", *Journal of Monetary Economics,* 2:1-32.

Dyson, F.D., 1988, *Infinite in All Directions,* Harper & Row, New York.

Fan, K., 1952, "Fixed-point and minimax theorem in locally convex topological linear spaces", *Proceedings of the National Academy of Science,* 38:121-126.

Harsanyi, J.C., 1967-1968, "Games with incomplete information played by Bayesian players", *Management Science,* I, II, III, 14:159-182; 320-334; 486-502.

Lehmann, E.L., 1959, *Testing Statistical Hypothesis,* John Wiley & Sons, Inc., New York.

Lucas, R.E.Jr., 1978, "Asset prices in an exchange economy", *Econometrica,* 46:1429-1445.

Meister, H., 1990, *The Purification Problem for Constrained Games with Incomplete Information,* Lecture Notes, 295, Springer-Verlag, Heidelberg.

Mertens, J.-F. and S. Zamir, 1985, "Formulation of Bayesian analysis for games with incomplete information", *International Journal of Game Theory,* 14:1-29.

Milgrom, P. and R. Weber, 1985, "Distribution strategies for games with incomplete information", *Mathematics of Operations Research* 10:4:619-632.

Muth, J.F., 1961, "Rational expectations and the theory of price movements", *Econometrica,* 29:315-335.

Postlewaite, A. and D. Schmeidler, 1987, "Differential Information and Strategic Behavior in Economic Environments", in T. Groves, R. Radner and S. Reiter (eds.), *Information, Incentives & Economic Mechanisms,* Basil Blackwell, Oxford.

Radner, R. and R.W. Rosenthal, 1982, "Private information and pure-strategy equilibria", *Mathematics of Operations Research* 7:3:401-409.

Sargent, T.J., 1973, "Rational expectations, The real rate of interest and the natural rate of unemployment", *Brookings Papers on Economic Activity* 2:429-472.

Wieczorek, A., 1984, "Constrained and Indefinite Games and Their Applications", Institute of Computer Science, Polish Academy of Science, Walsaw.

CHAPTER 9

Entrepreneurship in the New Technological Regime

Norio Okada

9.1 INTRODUCTION

What is meant by entrepreneurship? Is it changing due to the new dominant technological regime which is fostering a cosmo-creative climate?

In the conventional sense of the term, as used by Schumpeter (1937), an enterpriser is an economic agent who functions to actively pursue the 'achievement of new combinations' in the form of an 'enterprise'. In order to establish a new enterprise, the enterpriser must promote innovations. Thus we may tentatively say that 'entrepreneurship' is the spirit of the enterpriser in the above sense.

As the world faces the Fourth Logistical Revolution, as Andersson (1985) describes it, the notion of entrepreneurship may be forced to grow and shift. The newly evolving technological regime is characterized by innovative forms of progress which are being made in the areas of computers and communications, bio-engineering, material engineering, and other high-tech areas. One consequence of this is the increasingly shorter product cycle and new trends which are emerging in manufacturing technology, thus forcing manufacturers to shift their goal from 'efficiency' to 'flexible design' (Wiseman, 1990). According to Wiseman, efficiency is marked by mass production and repetitive routine works, whereas flexible design is characterized by diversity in type with small scale production, a rapid response to customer demand and adaptive change under uncertainty.

The changing needs of society - due to changes in lifestyle, diverse human tastes, and the pursuit of a higher quality of life, etc. - and the newly-evolving technological regime both interact with each other like the two wheels of a cart. What then does this mean for a contemporary enterpriser who hopes to initiate a new scheme? What will 'entrepreneurship' come to imply in the context of the Fourth Logistical Revolution?

9.2 NETWORK ORGANIZATION

In response to the question posed above, we now turn our attention to the changing role of 'information'. In the following discussion the term 'information' includes 'knowledge'. In fact, the role and impact of information have been changing to such an extent that the changes may be viewed as a 'revolution'. Thanks to the remarkable ongoing advancements in communications and transportation, for example, physical distance has become a far less critical constraint on the dissemination of information. Drastic improvements have been made in relation to the reduction of the time and financial resources needed for the generation, accumulation and processing of information, let alone its dissemination. All this has brought about the increasing globalization of information.

The more society and industry avail themselves of this revolutionary change in information, the more open and flexible they will become. It also becomes much more easy and beneficial for a society or industry to connect itself with others. It is reasonable to say that the ongoing logistical revolution is characterized by the networking of societies and industries which largely benefit from so doing.

Referring to this fact, Miyazawa (1988) claims that stepping further beyond the age of an 'economy of scope', we are now entering the age of an 'economy of networking'. His argument is that economies of scope refer to the production factors common to, say, two activities; these can be transferred from one activity to another at a relatively low cost. Economies of networking mean more than that; assuming multiple (economic) agents to be involved, this concept implies that there will possibly be 'synergetic effects' to be expected when multiple agents merge interactively. Notably, the term 'multiple-agency' and 'synergy' are characteristics of an economy of networking.

Examining this from the perspective of information and communications as well as from that of organization, we may legitimately claim that there is a shift of focus being made from a hierarchical or vertical structure onto a horizontal or network structure. Figure 9.1 compares these two types of organizational structures. The former concentrates crucial information at the top and then conveys information from the top to direct those further down below. In contrast, the latter allows those who are at peripheral locations to act autonomously in the generation, accumulation, processing and dissemination of information to a reasonable extent.

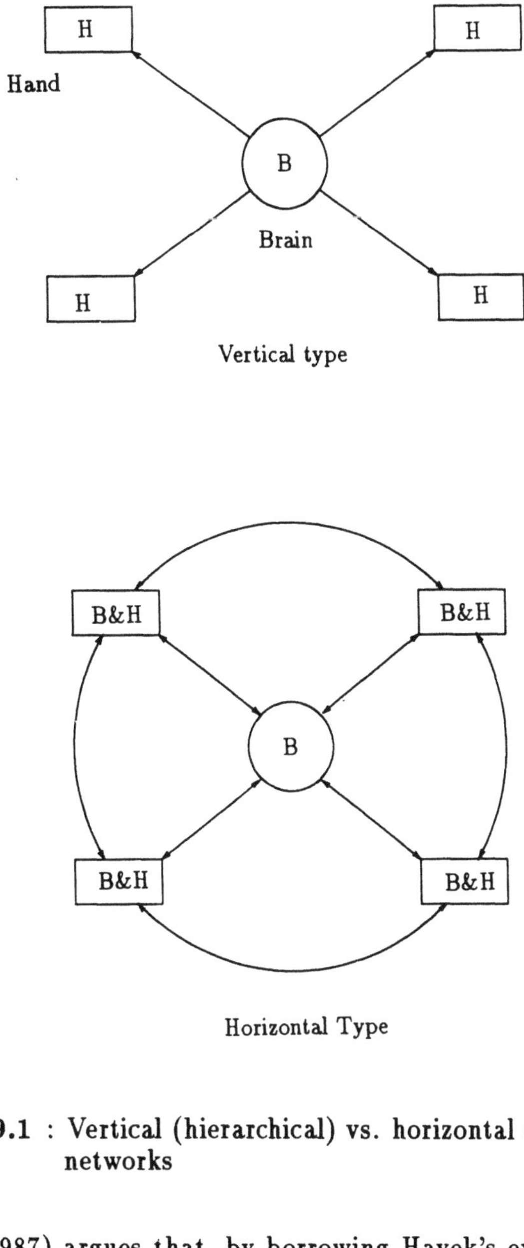

Figure 9.1 : Vertical (hierarchical) vs. horizontal (distributive) networks

Imai (1987) argues that, by borrowing Hayek's expression, information critical to the economy is owned by the 'man on the spot'; that is to say that 'on the spot information' must be reprocessed and reproduced incessantly at points which are located closest to the sources of information, so these local points have become increasingly vital to the economy. A network structure that is horizontal and autonomous may be better suited to the handling of 'on the spot information'.

As Okada (1990) points out, there is an analogy between this shift in organizational structure and the functioning of the brain. It is as if many of the 'brain's functions' have increasingly been required to be moved from the top or centre of an organization to its ends or 'hands'. In other words the ends of the organization are acquiring more of its brain-related functions and fewer of the functions of its hands that are totally subject to its brain (see Figure 9.1). He argues that due to the growing demand for customerization - as claimed by Wiseman (1990), amongst others - 'hands-on experience' and 'first-hand information' are becoming increasingly critical to successful industrial activity. This in turn leads us to the question: how should we encourage the society or industry concerned to reorganize itself into a network organization? What makes it possible?

Imai (1990) answers this by claiming that entrepreneurs, in the current sense of the term, constitute a set of economic agents who create a new context in society through networking. They make full use of this new context to combine those potential needs which are distributively embedded in society with existing resources which have not yet been well exploited. Thus, a market system which is closely knit into the network of economic agents is created. So entrepreneurship should be defined in the collective sense of a network organization, and not merely in the individual sense of an entrepreneur as was tentatively proposed in the introduction of this Chapter.

However, there is still another point to explain. There must be some factor that is attributable to the region in which economic activities are performed and that has a significant bearing on the manner in which networking is fostered as well as the extent to which entrepreneurship reveals itself in the region. What is this missing factor?

9.3 REGION AS A CREATIVE FORUM: COSMO CREATIVE CLIMATE

We now come to the pith of our discussion. We intend to introduce a concept which might be likened to the notion of a 'field', as in a magnetic field. Let us call it a 'cosmo-creative climate'. This term is used to describe the 'creative forum' which a region can set up to stimulate those potential economic agents who intend to establish a new enterprise in a network. It is a field which acts upon them, boosting them so that they can eventually succeed in undertaking a new enterprise. In other words, a cosmo-creative climate is a collective regional property that ferments entrepreneurship into the form of a network. Let us refer to this broadened sense of entrepreneurship as 'cosmo-creative entrepreneurship' or 'regional creativity'.

In fact, Okada and Kobayashi (1989) have studied this theme and used the term 'regional creativity', claiming that one should carefully differentiate between it and 'individual creativity'. They also introduced the notion of a 'region as a creative forum'. They argue that a cosmo-creative climate is the collective property of a 'region as a forum' in which the 'creativity of individuals' or the entrepreneurship of a collection of individuals are fostered into cosmo-creative entrepreneurship. The forum grows by absorbing those properties which have been produced through the materialization of cosmo-creative entrepreneurship into itself. The creativity of a region refers to its capacity or facility to foster the fermentation of new ideas or technology so that they may eventually reach the phase of product adoption and diffusion. Characteristically, a capacity of this kind can be dynamic in nature and it can grow by absorbing the gains of earlier rounds of innovations implemented in the region into itself.

In the following we will elaborate on this point and construct a conceptual model. We propose that the region be viewed as a forum in which players acting as decision-makers interact and collaborate in cultivating the seeds of invention, thus promoting a superior form of cosmo-creative entrepreneurship in the region. The model as a whole is a communication game of entrepreneurship (an entrepreneurial game) performed by players.

9.4 ENTREPRENEURIAL GAME

9.4.1 Model Formulation

According to Okada and Kobayashi (1989), it is assumed that there are nine players in this game. They are the initiator (I), comrade (C), appreciator (A), director (D), technical supporter (T), circulator (CR), finanacier (F), user (U) and imitator (IM). Each player is an individual or group who is assumed to be an independent decision-maker. However, in practice various players' roles could be performed by the same individual or group. The game takes the form of communication interplays performed by players. The players start the game by interacting with one another to form and expand their own communication network. A description of the players follows.

Initiator

This is a creative individual or group who invents some novel idea or new form of technology. He is assumed to initiate this entrepreneurial game by arousing awareness about his invention. In reality, it may not always be the case that an inventor becomes the initiator of the game. As a result, we will immediately modify the above assumption about the initiator. He may not necessarily be the inventor of the novel idea or new form of technology.

Comrade

The comrade is an individual or group who shares the goal of bringing the invented idea or new form of technology into being and assists the initiator in this effort.

Appreciator

The appreciator is an individual or group who literally appreciates the value of the invention. This player performs a critical role in the game. As well as appreciating the value and potential impact of what has been invented, this player is also authoritative enough to exert a far-reaching influence on other players so that they accept his appreciation of the invention.

Director

The director is an individual or group who literally directs the game so that the invented idea or new form of technology may finally be adopted and so that - if possible - the region may harvest the fruits of adoption (such as increased job opportunities). The essential role of this player is to act as a top manager who issues directions to resolve conflicts among the players, and to reduce the risks involved in the non-routinized novel enterprise.

Technical Supporter

The technical supporter is an engineering person or group who stays outside the core team formed by the initiator, comrade and director. This player supports the core team by offering them any vitally important engineering skills which they do not possess.

Circulator

The circulator is an individual or group who literally circulates the information about the invention among the players. He plays the critical role of disseminating the news about the invention among the players, as well as replaying the players' own assessments of the invention. It is noted that in order to achieve his task the circulator must have an extensive network of his own for information and communications.

Financier

The financier is an individual or group who offers financial assistance to the initiator or to the core team. Raising funds is a critical factor which determines the success of creative development.

User

The user is an individual or group who makes a decision about whether or not to adopt the creative enterprise as a novel product. This player performs an important role in judging whether the novel product truly meets his potential needs.

Imitator

The imitator is an individual or group who starts to imitate the novel

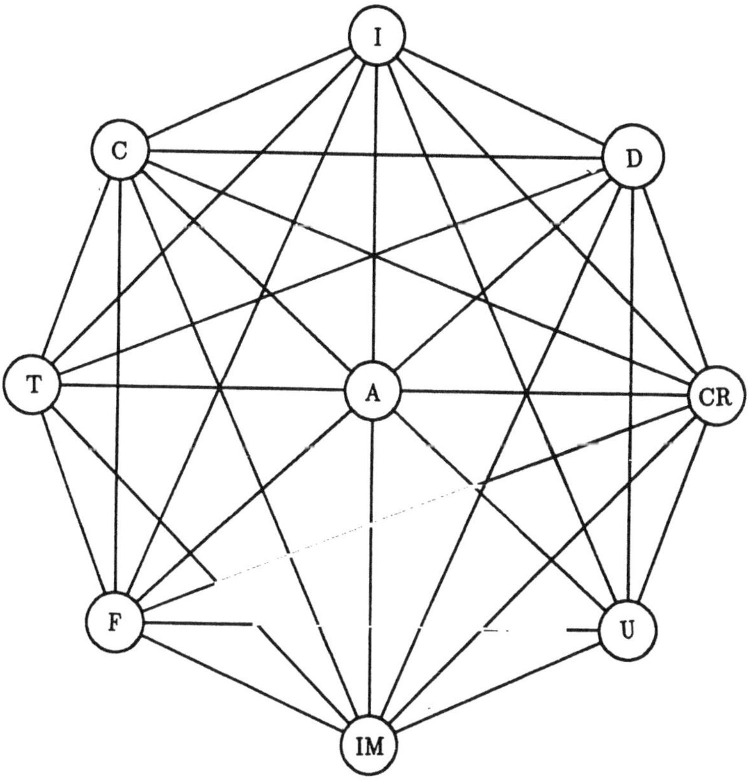

Figure 9.2 : General skeleton of the communication network model

product by gaining information about it and then producing the technology necessary for duplicating the product.

Figure 9.2 illustrates the structure of the game viewed as a set of communication interplays performed by the above-defined nine players. Table 9.1 shows how the commitment of the players can be categorized during the different phases of innovation.

9.4.2 Case Study of *Hyo-on* Research Institute

(1) Problem description
The *Hyo-on* Research Institute, which was established in 1985, is one of the leading creative enterprises in the Prefecture of Tottori, located on the Sea of Japan, about 200 kilometers northwest of Osaka. The Institute is situated in Yonago City, with a population of about 140,000. The president of this research institute is Dr. Yamane. He first in-

Table 9.1 : Matrix representing phases of participation in innovation

players	pre-invention	pre-adoption								post-adoption		
	embryogeny	invention	awareness	interest	study	trials	adoption	adaptation	implementation	diffusion	obsolescence	decay
I												
C												
A												
D												
T												
CR												
F												
U												
IM												

vented what is now well known to the world as a revolutionary form of technology. This was initially developed in relation to the cold storage of foods, but has since been extensively applied not only to the food and distribution industries but also to medical engineering, bioengineering, etc. Yamane gave the newly-developed technology the name of *hyo-on*, literally meaning 'icily warming' or 'icily tempering' in Chinese characters.

As Yamane explains, *hyo-on* is not simply the name of a new cold-storage technique, but may potentially be an extremely fundamental concept which could evolve into a new engineering and scientific discipline. For the sake of simplicity, however, it may suffice to perceive it approximately as a creative invention which focuses on a delicate temperature zone that causes any biological object to become frozen to dormancy. The term *hyo-on* refers to this temperature zone with the implication that the biological object is harshly (icily) trained (tempered) to an extreme state of dormancy, under which state its biological viability can be conserved to a maximum extent. This creative technology has currently been finding extensive applications in unexpectedly broad areas ranging from food storage and distribution in the service industry, to conservation and transplantation in medical science.

After graduation from what is now the Faculty of Agriculture of Tottori University, Yamane began research work at the Research Institute of Food Processing attached to Tottori Prefecture. There he became exposed to the body of knowledge and experiences concerning improvements in the breeding of the Tottori pear, a well-reputed and commercially successful local product which was first developed by the university and the prefectural institute. The pear is a delicate fruit which loses its freshness very quickly and so requires a sophisticated technology for storage and conservation. Yamane's discovery of the *hyo-on* technology in the early 70's was a result of his long-term efforts to invent a form of technology to best conserve and store the Tottori pear.

He first officially publicized this achievement in 1971, during a meeting at which engineers, practitioners and government officials of the Ministry of Agriculture gathered to assess newly-developed techniques of food processing and distribution. There he claimed that what he had developed was a creative technology which had no equal in the conventional field of food storage conservation. He insisted that the novel technique needed a new name and that it should be called *hyo-on*.

His claim was not immediately accepted by most of the participants, who came from different regions of Japan, including Tottori. A high government official - a relatively young member of the elite who was acting as a moderator at his session - was the only one to appreciate the potential value of *hyo-on* technology. He finally persuaded the other participants to accept his assessment. This was practically the first occasion upon which appreciation for Yamane's work had been officially expressed by any person or organization outside of his affiliated one. Thus, the news about his invention and the appreciation of an authoritative outsider was first circulated by word of mouth.

However, another decade passed before he received official academic appreciation of his work. A scientific award was offered by the Japan Society of Food Science in 1982; this was immediately followed by a call for an invited lecture at his Award Memorial Meeting organized by the Food Association of Tottori Prefecture. Yamane then received enormous attention from the mass media, who started to report his achievement across the country via newspapers and broadcasting networks.

By this time he had been contacted by large electrical manufacturing companies as a result of his own personal networks as well as that of the then governor of Tottori. Among them, a Tottori-based firm (SANYO) succeeded in introducing his idea into a new refrigerator model incorporating a *hyo-on* freezing box. The TV commercials which advertised this new model by frequently repeating the word *hyo-on*, contributed greatly to the popularization of this new term, as well as the refrigerator's commercial success.

Everything seemed to be synchronized. Yamane retired from the prefectural research institute as Director in 1983. He was approached by the Mitsubishi Trading Company, Nichimen Jitsugyo Trading Company, Japan Mail Steamer Company (Nihon Yusen), Japan Airlines

Table 9.2 : List of players

Initiator (I)	Dr. Yamane
Comrades (C)	Staff of Prefectural Institute
Appreciator (1) (A_1) Appreciator (2) (A_2) Appreciator (3) (A_3)	Government Official Scientific Community (Food and Agriculture) Scientific Community (Medical Science)
Director	Governor
Technical Supporter (T)	None
Circulator (1) (CR_1) Circulator (2) (CR_2)	Anonymous Mass Media
Financier (F)	Outside banks and firms (majority) Local banks and firms (minority)
User (U)	Outside big firms (majority) Local small firms (minority)
Imitators (IM)	None (still patent-bound)

and many other leading national companies. This led to two new developments:
(i) Yamane set up a new privately-run research institute as a company (joint stock corporation) in 1983, with a view to further developing and disseminating what he had come to conceive as the pioneering scientific and engineering discipline of *hyo-on*.
(ii) The Japanese distribution system also experienced remarkable innovations in 'chilled-storage and transportation' due to the introduction of this novel form of technology.

Innovations took place in surface, seaborne and airborne transportation, both within Japan and throughout the world. In the meantime the local food industry - including producers of pickles, soy-bean-cakes (tofu) and noodles - as well as its related service industry started to show interest in *hyo-on* and some producers even began to produce test products of their own and to sell them at local airports, department stores and retail shops.

Another stream of development is currently progressing in medical science. The technology of *hyo-on* has caught the attention of medical scientists. The idea is to apply this novel form of technology to the

Table 9.3 : Phases of participation in the innovation

players	pre-invention	pre-adoption							
	embry-ogeny	inven-tion	aware-ness	inter-est	study	trails	adop-tion	adap-tation	implemen-tation
I	**	**	**	**	**	**	**	**	**
C			**	**	**	**	**	**	**
A	(**)		*	**	**	**	**	**	
D	**								
T									
CR				*	*	**	**	**	**
F						(*)	*	**	**
U						(*)	**	**	**
IM									

(Note) * indicates that the corresponding player enters in this phase
** indicates that the player belongs to the region

fields of storing, conserving and transporting living organisms such as human internal organs. Collaborative research and experiments are proceeding between Yamane's research institute and the school of medicine of a Tokyo-based university.

The newly-built Research Institute of *Hyo-on* is managed in the form of a joint stock corporation with the above-mentioned companies as well as some leading nationwide and local banks who participate in the management as its stock holders and financiers.

Clearly the region of Tottori Prefecture and its neighboring areas have successfully started to enjoy the benefits of this invention to the extent that it has been brought into the phase of product adoption. However, a major problem still remains. The form of participation by local establishments has so far been largely 'passive' in the sense that (i) there has virtually been no technical support from the local industries, (ii) a 'hands-off' or 'wait-and-see' approach has characterized the attitudes of the management of local industries to the game, (iii) they have not actively expressed any appreciation, and (iv) only a small number of establishments outside the region have begun to use the invention. Can we make their attitudes more positive? Is this necessary? If so, how can we achieve this? Such questions may not be so easy to answer. We will, however, make a modest attempt to analyze the problem in a more systematic manner, by the use of the conceptual model represented in the previous Section.

(b) Model Application

Let us look upon the above-stated example as an entrepreneurial game. Table 9.2 lists the principal players who correspond to a particular individual, group, or organization. Given the above-described context, Table 9.3 fills in the cells of the matrix of Table 9.1. Figure 9.3 illustrates the skeleton of the structure of the game. The shaded nodes

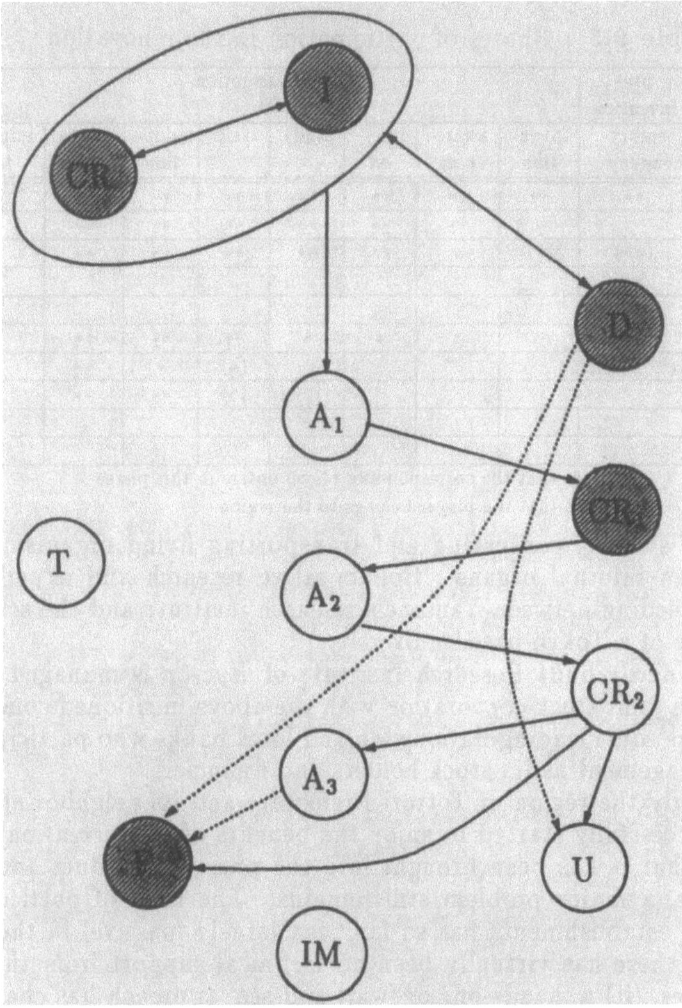

(N.B.) The shaded nodes indicats that they are partially or entirely part of the region.

Figure 9.3 : Growth of a communication network modeled for innovation of *hyo-on* technology

indicate that the players are partially or entirely from the region.

We immediately understand from Figure 9.3 that those who come from the local region are confined to the initiator, his comrades, and some of the circulators. Notably, none of the appreciators are from the region.

We then observe that the appreciators take a key role in this game.

They stay in the centre of the network, producing value-added information and upgrading the perceived level of the potential source (invention) so that their remarked information can be incited to flow to other players.

Let us recall that the region of Tottori itself failed to find local appreciators for the invention of *hyo-on* technology. To be fair, we immediately add that there seemed to have been some who did sense that the new achievement by Yamane could become something great. Yet this does not mean that they were 'appreciators' according to our definition. The difficulty experienced by local regions in finding appreciators of their own creates a barrier for local enterprisers and this is a common story for almost all local regions other than metropolitan ones.

However, we should underline that there is no particular reason whatsoever for local enterprisers to stick to their internal region as they endeavor to find appreciators for their achievements. That is to say that if the enterprise is open and flexible enough to extend its network beyond the region, this will provide a necessary, if not sufficient, condition for the region to foster a cosmo-creative entrepreneurship.

We now turn to a discussion of the circulators, who disseminate information about the invention, adding their own interpretation, developing illustrations and over- or understatements. Interestingly, circulators and appreciators tend to collaborate with each other. Circulators relay the appreciators' assessments to other appreciators or users.

Financiers enter the game at a relatively late stage. They naturally tend to take a wait-and-see, risk-aversive policy - especially those who come from the region. Appreciators and circulators who collaborate with the director perform critical roles in inviting them to the game.

There is virtually no player who provides any technical support (since they are represented by an isolated node.) So far no imitator has emerged because the technology is protected by a patent. Therefore, this does not mean that it has not been well diffused.

There exist players who correspond to local users but the number of such players is very limited. This entrepreneurial game seemingly owes much of its performance to areas outside the region such as Tokyo and Osaka and less to the region itself. This suggests that the region has a relatively small capacity (receptability) to bring an individual invention to the phase of harvesting its fruits.

Another way to look at the problem is that in the process of the invented technology finding its own users, those players who stay outside the region seem to take essential roles. We may interpret this such that they are in a more advantageous position than those potential users who stay inside the region; external players can take a more objective view of the value of the invention. Yet there seems to be another factor involved. As mentioned in Section 9.2, physical distance among players may no longer be a critical problem. The less critical physical distance becomes, the more critical cognitive distance among them becomes as a matter of relativity. In this sense those regions which serve as both sources and sinks of information of all kinds are placed in a more advantageous position since they can more easily reduce their cognitive distance from other regions in situations where vital information is produced and stored behind the scenes. In Japan, the metropolitan regions of Tokyo, Osaka, etc., are regions of this type.

As a matter of fact, in the beginning of the process of technology diffusion, users tend to be propagated in a highly discrete, non-spatial diffusive manner; in the due course of expanding their network they take a leap from the original region to another that enjoys a much higher level of information availability and appreciability. Generally we may expect that users are derived in a leaping process from one metropolitan region to another, while - in the meantime - they also start to diffuse within the source region. We found that this was precisely the case with our study. This fact again reconfirms our former statement that if the region is open and flexible enough, networking with the outside will interactively substitute for the roles of missing local players. Thus, cosmo-creative entrepreneurship would possibly be fostered.

9.5 CONCLUSION

We now conclude our discussions. The following are our major points.
(i) With the Fourth Logistical Revolution currently proceeding, entrepreneurship should be envisioned from the viewpoint of network formation with multiple economic agents involved.
(ii) From the viewpoint of regional development strategies, we should introduce a new concept. Namely, any region that encourages potential entrepreneurs to venture into a new project tends to be equipped with a collective regional property that ferments entrepreneurship into a network organization. We refer to the setting of the region installed with such a property as a 'cosmo-creative climate'.
(iii) We then proposed a conceptual model for analyzing the mechanism of network formation by multiple agents pursuing entrepreneurship. The model is a communication game of entrepreneurship with an interplay of players. A case study of Tottori Prefecture was conducted to demonstrate the applicability of the model.
(iv) It has been made clear through the application of this model that a local region like Tottori is handicapped in many ways. The difficulty

rests with linking initiators with appreciators, circulators, financiers, users, etc.
(v) However, it is important to note that if a local region is open and flexible enough to link its network with the outside, it may possibly overcome its handicaps. Such a condition is precisely intended to be conveyed by cosmo-creative entrepreneurship - entrepreneurship in the new technological regime.

One question still remains: how to foster a cosmo-creative climate? All we can say so far is that thanks to various remarkable advancements, the situation has become favorable for local regions, assuming that the invention at stake is vital enough to penetrate the boundaries of the region to the outside. However, how can we promote this? To answer this question, we still need to conduct further research.

REFERENCES

Andersson, Å.E., 1985, *Creativity: The Future of the Metropolis*, Prisma, Stockholm (in Swedish).

Imai, K., 1987, "Information Economics and Information Edition - Helmes's Perspective of Entrepreneurship", in K. Imai, et al. (eds.), *Ecology of Economics*, NTT Publications, Tokyo, pp. 17-18 (in Japanese).

Miyazawa, K., 1988, *Cross-Industrialization and Information Ages - Their Impacts on Industry and Society*, Yuhikaku Publ. Comp., Tokyo, pp. 54-60 (in Japanese).

Okada. N., and K. Kobayashi, 1989, "Region as a Creative Forum: A Conceptual Approach", CWP-1989:23, CERUM, UmeåUniversity, Sweden.

Okada, N., 1990, "R & D Oriented Regional Policy: Perspective and Implications", *Proceedings of the International Forum on Logistical Development and Its Regional Consequences in Osaka - Towards a Cosmo-Creative City*, pp.105-112, Osaka (in Japanese).

Schumpeter, J.A., 1937, *The Theory of Economic Development*, Harvard University Press, Cambridge, MA.

Wiseman, M., 1990, "Product, Process, and Strategy in Knowledge Based Regional Development Policy", *Proceedings of the International Forum on Logistical Development and its Regional Consequences in Osaka - Towards a Cosmo-Creative City*, Supl., Osaka.

CHAPTER 10

Network Cities Versus Central Place Cities: Building a Cosmo-Creative Constellation

David F. Batten

10.1 INTRODUCTION

In today's global network economy, knowledge corridors link major knowledge-intensive nodes or hubs to form network cities. European examples include the Cambridge-London and Stockholm-Uppsala corridors as well as Randstad Holland. The Kansai region of Japan is another example of a network city. Analytical tools for the analysis of knowledge exchange processes between different centers are also under development (see, among others, Batten, Kobayashi and Andersson, 1989; Batten and Törnqvist, 1990; Kobayashi, Batten and Andersson, 1991).

Such urban corridors possesses an intrinsic network character which may be contrasted with those unicentric metropolises where creative activity is also generated. Since networking is feasible between centers of various sizes, it may offer medium-sized cities an alternative long-run strategy to the more daunting prospect of competition with all their neighbours.

This Chapter builds an empirical case for network cities and outlines a simple analytical framework to compare them with unicentric cities. The basic model is crude and rather unrealistic at this stage, but may easily be refined to encompass additional factors. Key features of the model are:

(1) that it regards population and infrastructure as the primary determinants of economic output in the long run;

Table 10.1a : The ten largest cities in Europe. 1000-2000
(by population in thousands)

1000		1400		1700		1900		2000	
City	Population	City	Population	City	Population	City	Population	City	Population
Constantinople	450	Paris	275	Constantinople	700	London	6480	Moscow	9000
Cordoba	450	Milan	125	London	550	Paris	3330	Paris	8500
Seville	90	Bruges	125	Paris	530	Berlin	2424	London	7000
Palermo	75	Venice	110	Naples	207	Vienna	1662	Leningrad	5500
Kiev	45	Granada	100	Lisbon	188	Leningrad	1439	Madrid	3000
Venice	45	Genoa	100	Amsterdam	172	Manchester	1255	Berlin	3000
Regensburg	40	Prague	95	Rome	149	Birmingham	1248	Rome	2800
Thessalonika	40	Rouen	70	Venice	144	Moscow	1120	Birmingham	2500
Amalfi	35	Seville	70	Moscow	130	Glasgow	1072	Manchester	2500
Rome	35	Ghent	70	Milan	124	Liverpool	940	Budapest	2100

Sources : Hohenberg and Lees (1985) and author's estimates.

(2) that it emphasizes their qualitative as well as their quantitative dimensions; and

(3) that it allows for a stage of increasing and a stage of decreasing returns to urban economic scale.

10.2 THE LAW OF RISE AND FALL

Cities do rise and fall, albeit slowly in some cases. They possess vintage properties, in the sense that large and previously prosperous ones eventually decline in the absence of sufficient renewal and revitalization activity. The rise and fall of cities is indeed well documented in all parts of the world. To catch a glimpse of this lifecycle phenomena, we may turn to Europe over the last millennium (see Table 10.1a) or to the changing global hierarchy of cities over a much shorter period (see Table 10.1b). Many cities which dominated several centuries ago are relatively small today. Others, like Mexico City and Sao Paulo, have grown remarkably in the space of just fifty years.
Observation 1: Any model designed to explore the long-run prospects for city systems must allow for stages of rise and fall, i.e. stages of increasing and stages of decreasing returns to scale.

Table 10.1b : The world's ten largest urban areas, 1950 & 2000 (by population in thousands)

		Population (thousands) 1950		*Population (thousands) 2000*
1	New York-Northeastern New Jersey	12,300	Mexico City	31,000
2	London	10,400	São Paulo	25,800
3	Rhine-Ruhr	6,900	Tokyo-Yokohama	24,200
4	Tokyo-Yokohama	6,700	New York-Northeastern New Jersey	22,800
5	Shanghai	5,800	Shanghai	22,700
6	Paris	5,500	Peking	19,900
7	Greater Buenos Aires	5,300	Rio de Janeiro	19,000
8	Chicago-Northwestern Indiana	49,00	Greater Bombay	17,100
9	Moscow	4,800	Calcutta	16,700
10	Calcutta	4,400	Jakarta	16,600
11	Los Angeles -Long Beach	4,000	Seoul	14,200
12	Osaka	3,800	Los Angeles -Long Beach	14,200

Source : United Nations, *Patterns of Urban and Rural Population Growth* (*Population Studies*, 68), New York:United Nations, 1980.

10.3 NETWORK CITIES VERSUS CENTRAL PLACE CITIES

It is often argued that city size fosters innovative propensity. In the USA this has even been given a title: "The HVLT hypothesis" (after Hoover, Vernon, Lichtenburg and Thompson), since its authors emphasized the unique advantages enjoyed by New York City in the production of those commodities for which continual innovation or a constant flow of new information played an important role (see Hoover and Vernon, 1959). Whilst the supportive evidence linking size and innovative capacity is impressive, some recent observations suggest that part of the innovative growth potential which traditionally resided in larger U.S. and European cities may now be found in smaller urban concentrations (see, for example, Norton and Rees, 1979; Hohenberg and Lees, 1985). The innovative activity of multinational companies can be seeded in various locations simultaneously, and is by no means restricted to the creative resources of the big cities. Table 10.1 also reminds us that size alone has rarely been an adequate guarantee against eventual decline and obsolescence.

Such a trend reversal is consistent with the ongoing march towards a global network economy. The internationalization process implies a weakening in the relative importance of intraregional accessibility in

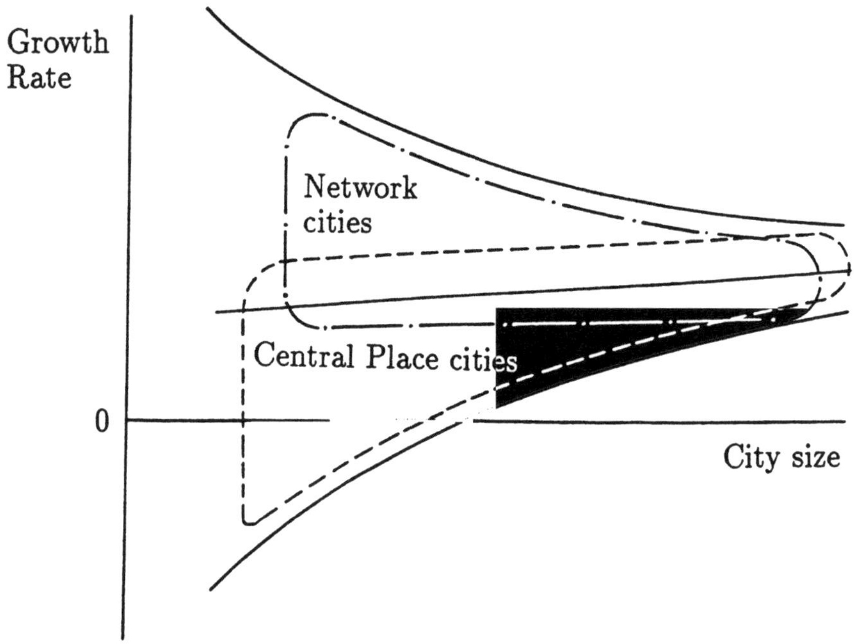

Figure 10.1 : Growth versus size in two systems of cities

favor of stronger international contacts. Mushrooming growth among firms responsible for information-processing , telecommunications and air transport capacity is greatly facilitating point-to-point contacts between many dispersed locations, thereby increasing the network character of the world economy. Knowledge about important innovations can spread rather quickly under these conditions. As these tendencies proliferate further, the geographical contiguity of regions and the relative size of a place in a local context may become less important than they were in the past. The urban-to-rural gap is widening. Today we must concern ourselves with diverse changes permeating across interdependent networks. In this new arena, a global network perspective is mandatory.

In the traditional theory of central places, growth potential is proportional to size. This need not be the case for a network city. It is in fact network cities and urban "hubs" that have accounted for an above-average share of all urban development in today's Europe (see Figure 10.1). Although some larger cities and urban hubs possess both network and central place characteristics, it is the smaller network cities that have counteracted the central place trend towards

Table 10.2 : The ten most important creative regions of Europe

	(NETWORK) CITY	SCIENTIFIC CREATIVITY	AIR TRANSPORT CAPACITY
1	London-Cambridge-Oxford	100	100
2	Paris	70	69
3	Randstad-Holland	33	25
4	Bonn-Dusseldorf-Cologne	33	21
5	Stockholm-Uppsala	26	23
6	Brussels-Louvain-Ghent	24	12
7	Frankfurt-Mainz-Giessen	22	42
8	Munich	22	17
9	Heidelberg-Karlsruhe-Stuttgart	22	17
10	Copenhagen	19	21

Sources : Science and Social Science Indexes, INFORSK, Umeå, Sweden; Air Transport Statistics, 1988; Batten and Johansson,1991.

primacy and contributed to the **size-neutrality** of urban growth (see Robinson, 1973; Hohenberg and Lee, 1985).

The natural extension of this trend is that future phases of growth and decline among systems of cities may be intimately associated with each city's acquired role in both national and international economies. It also suggests that competitive leadership may not be restricted to those few larger unicentric cities which have served traditionally as seedbeds of innovative activity. Smaller settlements may be able to enjoy relative prosperity by specializing in specific innovations which capitalize upon their "interactivity" across associated networks. Knowledge-sharing becomes a more realizable process. It is worthwhile noting that seven of the ten most creative European regions of today are more-or-less network cities (see Table 10.2).

Observation 2: Network cities are assuming greater importance as the scope and scale of economic activity extends into the international domain and the need for creative resources increases.

10.4 TODAY'S NETWORK ECONOMY OF EUROPE

The seeds of a new European network economy were sown as far back as the 11th century, when safer trade routes triggered the revival of

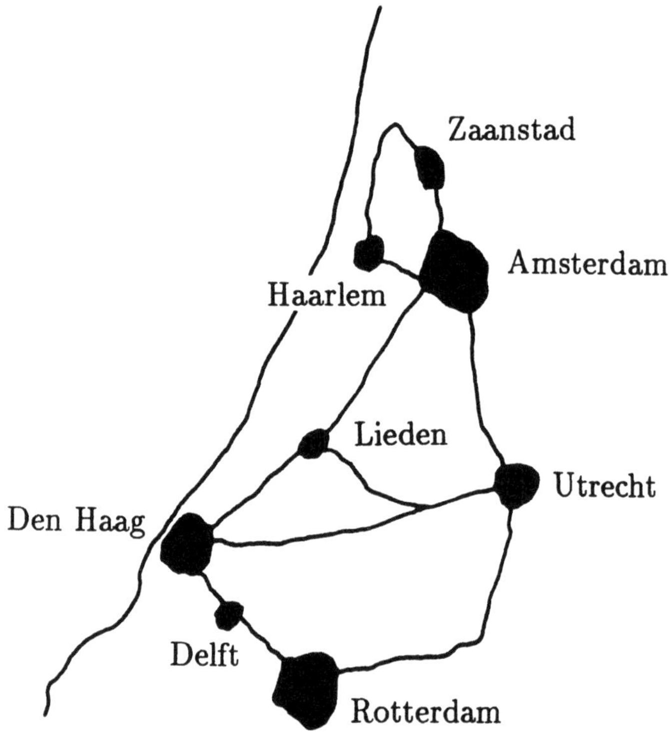

Figure 10.2 : A modern network economy: Randstad Holland

many medieval cities (Andersson, 1985; Batten and Johansson, 1991). To see this network economy evolving into its modern equivalent, we may turn to the Netherlands in the 19th century. The new economic stimulus of development along the Rhine and its tributaries promoted the rapid growth of two Dutch cities to challenge Amsterdam's primacy - The Hague as the royal residence and political capital and Rotterdam as the port commanding the mouth of the Rhine. Rotterdam and The Hague grew from 26% and 19% of Amsterdam's size, respectively, in 1800 to 102% and 67% by 1985. Perhaps more significantly, the distance between each of them shortened appreciably. Other smaller centers, such as Delft, Leiden, Haarlem and Utrecht, also grew and spread. Today the cities and towns of Holland have converged so closely to each other that they form a unique network city. Having the shape of a giant horseshoe, the Dutch choose to call it Randstad Holland: the Ring City. With Schiphol airport at its center, this city system is one of the most accessible urban agglomerations in the world.

The network pattern is possibly clearer in Holland because its land has been so actively managed. A similar network city is developing in the Kansai region of Japan, focusing on Osaka, Kyoto, and Kobe. In this type of network system, close links are forged between places with complementary functions rather than simply on the basis of distance or demand thresholds.

A network city may generally be viewed as a more sophisticated combination of urban corridors. The latter consists of two (or more) major nodes linked by efficient infrastructure. It may also include some smaller intervening nodes and is often associated with a science precinct including clusters of high-technology firms. European examples of this type of corridor development include the M11 Corridor linking Cambridge and London, as well as the Stockholm-Uppsala corridor. Each of these corridors also incorporates an international airport. Since the airport-university combination turns out to be one of the most synergistic factors currently contributing to faster and more prosperous urban growth in the Swedish context (Andersson, Anderstig and Hårsman, 1987), those highly accessible places nurturing a higher level of knowledge-based activities may be expected to prosper throughout Europe in relative terms. These places may well form tomorrow's system of cosmo-creative networks, being members of the C-Society articulated by Andersson (1985).

Observation 3: There is a need to develop an analytical framework which facilitates the comparison of network cities with the more traditional unicentric and polycentric cities.

10.5 COMPARING UNICENTRIC AND NETWORK CITIES

In this section, a rudimentary comparative framework will be outlined. As a first step towards an eventual tool for comparative analysis, it captures some key aspects of the descriptive discussion outlined above. Note that the present version of the model is rather unrealistic and that the choice of variables is motivated primarily by a future objective: to describe the relatively slow processes of adjustment across those networks which form the substructure or "arena" on which other faster change processes take place. In a dynamic version of the static model sketched below, such a differentiation of variables and speeds of change will be essential in order to fully comprehend our chosen characterization of urban change. Let

$$X = a\{b(P+G)^2 - P^3 - G^3\} \qquad (1)$$

where $P = z(P^* + D), G = z'G^*$, and X measures the level of economic activity, P^* measures the size of the city's economically active population (workers and customers), D measures the size of its external customer network, G^* measures the city's stock of infrastructure,

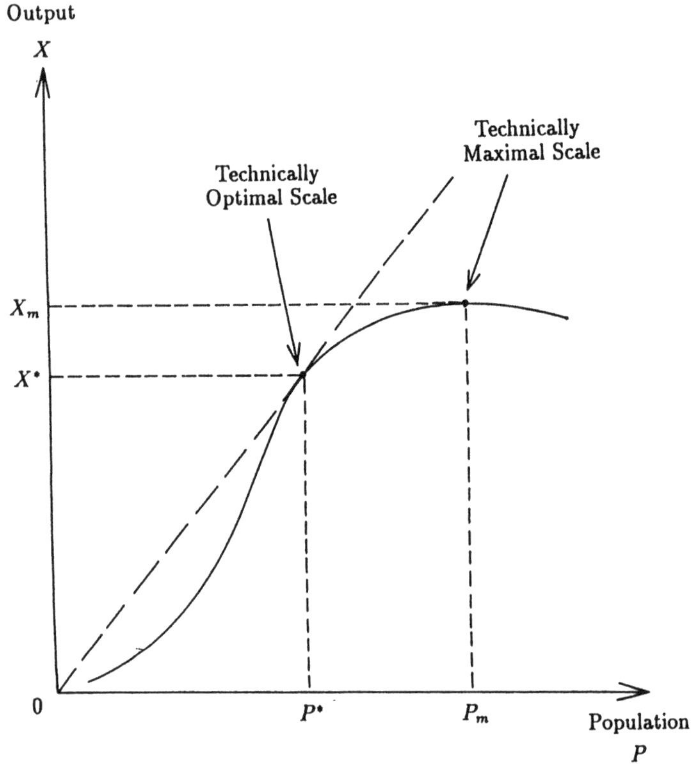

Figure 10.3 : Variable returns to scale in the urban lifecycle

a is an index of technological efficiency, b is an index of the scale of economic activity, z is an index of human skills, and z' is an index of infrastructure quality.

In the above production function for a city, D is intended to reflect the role (if any) of the city as an international market center. The size of its customer network should be measured on a relative scale of international links in comparison with local and regional links. Thus it might be weighted according to the ratio of international to regional economic activity.

Economic activity X reaches its maximum when both marginal productivities are zero, which occurs when

$$P = G = 4b/3. \tag{2}$$

The technically optimal scale is achieved when marginal and average productivities are equal, i.e. when

$$(P^3 + G^3)/(P+G)^2 = b/2, \tag{3}$$

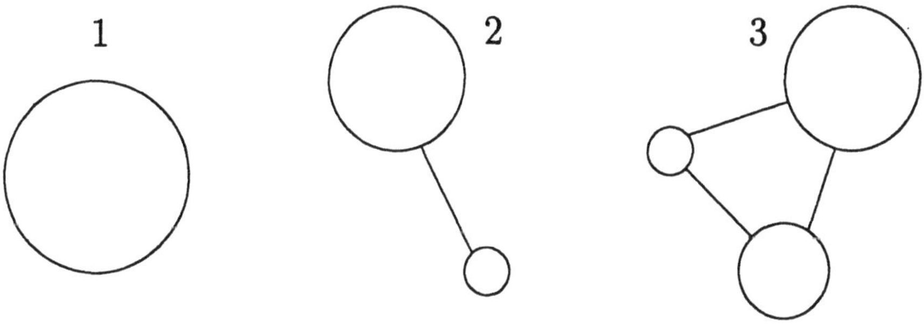

Figure 10.4 : Three types of urban configurations

and the term $2PG$ accounts for synergistic interactions between the local population, external customers and infrastructure.

It is illustrative to consider a situation in which $G = 0$, which might correspond to a simple agrarian economy. Then the model reduces to the following:

$$X = a\{bP^2 - P^3\}. \tag{4}$$

This function is plotted in Figure 10.3. It satisfies the "regular ultra passum law" proposed by Frisch (1965), in that it contains a zone of increasing returns to scale and a zone of decreasing returns to scale. Puu (1985) suggested this functional form as a modification to Hotelling's (1921) population growth model. In this simplified economy, the technically maximal scale occurs when $P = 2b/3$.

We can use the simplified version of the model - as given in (4) - for a comparative analysis of three types of urban systems: unicentric cities, bicentric urban corridors and polycentric urban networks. For simplicity, we shall restrict our present example to the unicentric, bicentric and tricentric cases (see Figure 10.4).

Referring again to Figure 10.3, we shall describe the stage of life-cycle of an urban system by its position along an $X - P$ curve. This may be approximated by its position within one of the following three zones of productivity:

(1) Increasing returns to scale: $P < b/2$;

(2) Decreasing returns to scale: $b/2 < P < 2b/3$;

(3) Decreasing output: $P > 2b/3$.

By assuming that $D = 0$ in this simplified agrarian economy, we can restrict our initial attention to the quality of the workforce (i.e. population), since we know that the effective population $P = zP^*$. To allow simple comparative analyses, we further assume that

(1) commuting distances within and between the nodes are insignificant;

(2) congestion is restricted to within the nodes rather than between them;

(3) the attractiveness of different urban configurations or sizes is not considered here, only their output.

Each of these restrictions may be relaxed at a later date to permit a more realistic analysis.

Our aim is to compare outputs X_1,X_2,X_3 (1 = unicentric, 2 = bicentric, 3 = tricentric city) under varying returns to scale, for equivalent populations

$$P_1 = P_2(1) + P_2(2) = P_3(1) + P_3(2) + P_3(3). \tag{5}$$

To simplify matters, the parameters a and b are held constant. The results are given in Figure 10.5, and are restricted to two different size configurations for each of the three classes of scale returns.

A few comparisons are perhaps worthy of note:

(1) When $P < b/2$, network cities with unequal node sizes are more productive than those with equal nodes;

(2) When $P = 2b/3$, bicentric urban corridors are just as productive as the unicentric city;

(3) When $P > 2b/3$, network cities are more productive than their unicentric counterparts, but those with unequal node sizes are inferior to those with equal nodes.

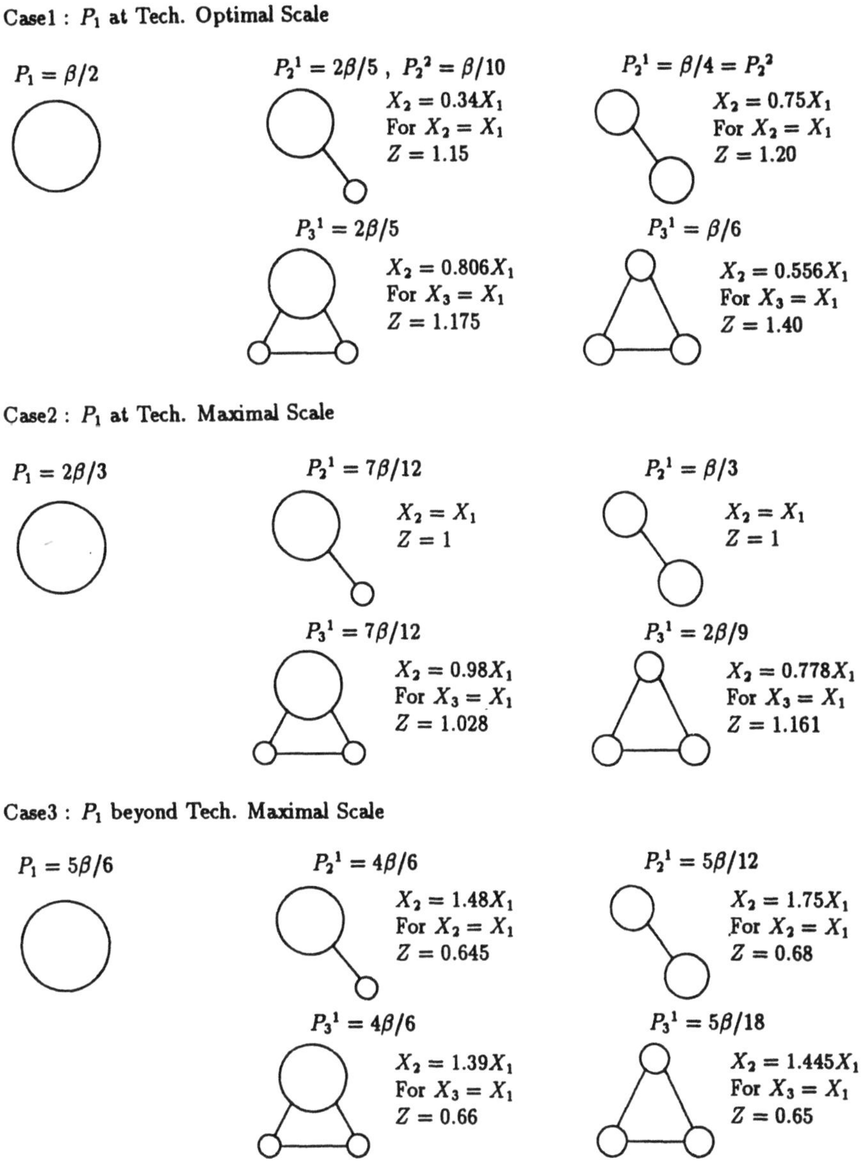

Figure 10.5 : Comparative analysis of three different urban configurations

It would be interesting to explore the effects of varying the external network variable D. In an agrarian economy, this might correspond to selling some crops to customers beyond your local region. To do this, however, improved means of transportation will be needed, beyond those utilized to deliver to local markets. Thus infrastructure investment is mandatory. We might therefore consider establishing a direct relationship between D and G. This task will be left for future investigation.

As D increases, the effective population P can rise without any change in P^*. The idea here is that some unicentric cities may be able to expand their scale of economic activity without necessarily using more local factors (population, capital, infrastructure). To be classified as a network city of this type, a threshold level of D/P activity would need to be reached. We shall refer to this urban system as a *Category I Network City.* Although unicentric, it possesses some of the features normally found in the multinodal configurations depicted in Figure 10.4. The latter may be described as *Category II Network Cities.*

10.6 CONCLUSIONS

In this short Chapter, our initial task was to demonstrate the growing importance of network cities - those bicentric urban corridors and multinodal urban networks which are playing a more influential role in today's global economy. Although our empirical evidence was brief, it did lend some support to this hypothesis.

To compare the attributes of network cities with unicentric ones, new analytical approaches are needed. One such approach was outlined. Although this model is crude and rather unrealistic at this stage, it may easily be refined to encompass additional factors. Key features of the model are (1) that it regards population and infrastructure as the primary determinants of economic output in the long run; (2) that it emphasizes their qualitative as well as their quantitative dimensions; and (3) that it allows for a stage of increasing and a stage of decreasing returns to urban economic scale.

It is too early to reach any definitive conclusions from this comparative model. Even so, our preliminary analyses suggest that opportunities may exist for medium-sized cities under conditions of increasing or decreasing scale returns:

(1) under an increasing returns regime, to strengthen their position in their central place hierarchy as well as to explore networking opportunities with larger cities;

(2) under a decreasing returns regime, to explore more vigorously networking opportunities with other larger and smaller settlements.

Both the past and the future of medium-sized cities may be viewed with stark simplicity in Figure 10.1. The path towards prosperity would seem to correspond closely with the path towards networking in its multifarious forms.

REFERENCES

Andersson, Å.E., 1985, *Creativity: The Future of the Metropolis*, Prisma, Stockholm (in Swedish).

Andersson, Å.E., Anderstig, C. and B. Hårsman, 1987, *"Knowledge and Communications Infrastructure and Regional Economic Change", CWP-87:25, CERUM, UmeåUniversity, Umeå.*

Batten, D.F. and B. Johansson, *1991, "Origins of the network economy", paper presented to the 31st European Congress of the Regional Science Association, Lisbon, August.*

Batten, D.F., Kobayashi, K. and Å.E. Andersson, *1989, "Knowledge, Nodes and Networks: An Analytical Perspective", in Å.E. Andersson, D.F. Batten and C. Karlsson (eds.), Knowledge and Industrial Organization, Springer-Verlag, Heidelberg, pp. 31-46.*

Batten, D.F. and G. Törnqvist, *1990, "Multilevel network barriers: The methodological challenge", The Annals of Regional Science, 24: 271-287.*

Frisch, R., *1965, Theory of Production, Reidel, Dordrecht.*

Hohenberg, P.M. and L.M. Lees, *1985, The Making of Urban Europe: 1000-1950, Harvard University Press, Cambridge, MA.*

Hoover E. and R. Vernon, *1959, Anatomy of a Metropolis, Harvard University Press, Cambridge, MA.*

Hotelling, H., *1921, "A Mathematical theory of migration", MA thesis presented to the University of Washington; republished in 1978 in Environment and Planning A, 10:1223-1239.*

Kobayashi, K., Batten, D.F. and Å.E. Andersson, *1991, "The sequential location of knowledge-oriented firms over time and space", Papers in Regional Science, 70:4:381-397.*

Norton R.D. and J. Rees, *1979, "The product cycle and the spatial decentralization of American manufacturing", Regional Studies, 13:141-151.*

Puu, T., *1985, "A simplified model of spatiotemporal population dynamics", Environment and Planning A, 17:1263-1269.*

Robinson, B.T., *1973, Urban Growth: An Approach, Methuen, London.*

United Nations, *1980, Patterns of Urban and Rural Population Growth, Population Studies 68, United Nations, New York.*

CHAPTER 11

Economic Evolution and Urban Infrastructure Dynamics

Börje Johansson

11.1 INFRASTRUCTURE AND ECONOMIC INTERACTION

11.1.1 Introduction

Infrastructure is often thought of as an arena for social and economic activities. The subsequent analysis considers infrastructure in the form of the built environment and associated networks, which function as systems for economic interaction. One central idea of this Chapter is that the development of product concepts, experimentation with young products, and routinised processing of mature and ageing products, all constitute distinct categories of activities such that each has its specific interaction characteristics and thus demands for particular combinations of infrastructure attributes of the built environment, in which it is located.

With this background the Chapter starts by outlining the features of urban infrastructure as an environment in which the attributes of products (goods and services) can develop and age, while the production activities relocate in a self-organised pattern. In the subsequent presentation the analysis strives to make precise the conditions behind the technological time arrow, which is implicit in the above type of spatial product cycle development.

A central theme in the described theoretical framework is the movements of activities between locations with different infrastructure attributes as new process vintages are gradually applied, and new products are introduced to replace established, mature product vintages.

A new element in this framework is introduced in Section 11.3, and consists of an explicit treatment of the ageing of infrastructure itself. By considering the interaction between economic activities and the features of the production infrastructure, a cyclic pattern of ageing and renewal of infrastructure is derived. A basic condition is the interplay between the very slow infrastructure investment adjustments and the relatively faster responses from the development and location/relocation of economic activities. The Chapter ends with empirical illustrations of the two fundamental dynamic processes analysed in Section 11.4.

The Chapter builds on and extends recent analyses in Andersson and Johansson (1984), Anderstig and Hårsman (1986), Johansson and Snickars (1988), Puu (1988), and Okada and Kobayashi (1989).

11.1.2 Capital and Infrastructure

Capital makes all the difference in intertemporal economics. Investment is generically a small fragment of total capital/capacity. Hence, in principle capital is always rationed in the short-term.

Capital has extension in time. Depending on the arrangement of the asset and pertinent institutional arrangements, the capital may be more or less public and have infrastructural properties in varying degrees. Infrastructure has a systemic nature and a spatial extension, and it constitutes a landscape or network of possibilities for interaction and communication in various ways (Currea and Polenske, 1985; Lakshmanan, 1989; Mera, 1984).

The above formulation corresponds to the idea that infrastructure is an arena for economic and social life. In the framework developed in this Chapter, the analysis concentrates on the built environment which exhibits infrastructure properties in varying degrees. These properties (attributes) include (i) capital with a fixed location and spatial extension, (ii) generality with regard to different categories of actors and types of activities, as well as over time, (iii) public good externalities in at least some dimension, and (iv) a systemic nature by providing interaction possibilities and by integrating activities around these possibilities. The above enumeration forms a vector of infrastructure attributes. Segments of a built environment will differ with regard to the amount of each vector element that they embody.

The main distinction is not between public and private capital, but between production-specific and production-general capital and aspects of the built environment. The former will lose its value when its production ends. In this sense the production-specific capital represents sunk costs. Infrastructure capital does not. This distinction between production-specific capital (including non-general buildings) and general infrastructure is reflected in sharp differences between their respective dynamic evolution and their influence on economic renewal processes.

11.1.3 Internodal and Intraurban Infrastructure

Infrastructure which facilitates interaction within urban regions and between urban nodes has the general character of networks. The long-term evolution of spatial economic systems is intrinsically determined by the interaction between the development of (i) intraurban (local) networks, and (ii) internodal nets of links. For example, new interaction networks within urban regions facilitate the further concentration of activities in every node. As a consequence the demand for interaction between two economically expanding nodes will be intensified, which in turn makes it profitable to invest in improved links between the nodes.

Intranodal networks which create infrastructure attributes of the built environment comprise water, sewerage, drainage, and energy supply systems, local road and rail systems, as well as different types of communication networks. These systems facilitate the production and distribution activities in urban nodes. In particular, they make it possible for economic actors to benefit from the nearness of dense urban structures. For trading and commercial city economies, the urban infrastructure has through history functioned as an arena on which producers, merchants, bankers and many other actors can establish, develop and restructure person-networks. The building and rebuilding of links between economic partners and between buyers and sellers constitute the most basic mechanism for the renewal and evolution of every market-like economy. These link-developing activities are almost exclusively carried out in urban environments with suitable infrastructure attributes (Johansson, 1989). This statement can be verified for the internationalised multinodal economies of today, as well as for the ancient Greek 'world economy' which was built on a network of urban economies in the eastern part of the Mediterranean region.

The link-formation described above is, in all essence, an investment in more or less durable channels for communication and the exchange of knowledge. The formation of these links requires personal contacts of a face-to-face type.

The same is also essentially true for the maintenance and active use of the channels. Interaction between persons is especially sensitive to logistical possibilities of coordinating the time-tables for individual contacts. When such nearby contacts are frequent the demand for appropriate urban infrastructure becomes accentuated. Such infrastructure combines accessibility in local networks with a dense environment

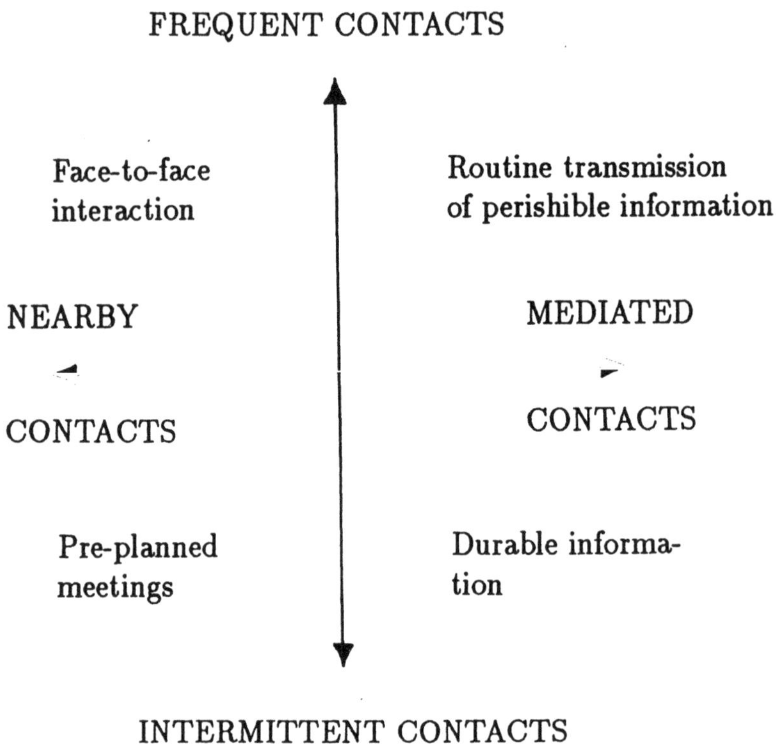

Figure 11.1 : Logistical requirements for person-interactions

of meeting places and multiple facilities, as well as a concentration of a variety of optional person-contacts, representing a multitude of experience, competence and skills.

Figure 11.1 presents a classification of contacts and interaction with regard to frequency and mediation form. The diagram contains four fields of contact and communication forms. In field (I) the demand for personal contacts is spread densely in time. The classical form of solving this problem is density in space, based on urban infrastructure. In field (IV) the interaction is concentrated to a few intervals in time. Field (II) refers to transmission by means of routine techniques of information which are perishable (like news). Field (III) represents transfers of information which are durable and may be indivisable. The development of communication techniques and logistical solutions determine how demand for interaction is satisfied by means of various combinations of the four principles. In the sequel the analysis concentrates on field (I), which is claimed to require intraurban acccessibility networks in dense structures.

11.1.4 Hotel Attributes of Urban Infrastructure

The metropolis and the urban region are infrastructure. They are a built system with infrastructure attributes, and their components and subsystems also have infrastructure properties. The above statement is not always self-evident. We can find a large variety of urban settlements which have been designed to serve very particular production activities. In such cases the urban environment should be classified as production-specific capital, based on investments without any generality. Urban nodes characterised by a strong export-orientation often have their entire structure designed in such a narrow sense. Some of them are known as company towns. But to provide a quite different example: Manchester city in the middle of the 19th century was an enormous textile factory with certain building accessories. Later in its life-cycle the city came to suffer greatly from its specialisation.

An urban region with a rich endowment of infrastructure attributes is general - not specific - in design. Then the fit between the physical characteristics and the economic processes in the region is not strict. Many of the buildings can be adapted easily to match varying purposes. The fact that the buildings are placed adjacent to one another in zonal complexes creates positive externalities. Concentration provides contact potentials which make it easy to interact at random preferences.

Urban processes can be categorised according to their speed of adjustment. For certain processes, individual actors make frequent decisions, and in such cases the processes often quickly reach a new equilibrium after a disturbance has occurred. A permanent disturbance brings about new conditions to which the decision makers have to adjust. Such changing conditions may also require new compositions of location patterns. When these changes require that the built environment is changed, the adjustment will in general constitute a slow process. It is in this context that we can fully appreciate the value of general, urban infrastructure. In a built environment with general and polyvalent characteristics, maturing activities may migrate away to other parts of an urban region, while new activities move in. Those production and other processes which can benefit from a certain location more than other activities, can squeeze the latter out from their original premises by offering higher rental payments. In this way the same built infrastructure can be used over and over again by new and alert economic actors. This means that the infrastructure capital does not represent sunk costs. The urban infrastructure may be said to display 'hotel attributes'. In the following Sections we argue that this hotel property stimulates a characteristic form of economic dynamics and is capable of speeding up economic renewal and evolution.

11.2 PRODUCT VINTAGES AND INFRASTRUCTURE REQUIREMENTS

11.2.1 Product and Process Vintages

According to the theory of product cycles, each product develops along a trajectory which often starts with an embryonic phase of research, laboratory testing, implementation of small scale production, and interaction with potential customers. In many cases a selected customer may cooperate in the development effort. As a rule, this experimental and *R&D*-oriented search activity is more frequent in certain types of economic environments than in others. This first search phase of a product cycle is intimately coupled with early imitation activities which elaborate and vary the theme established by the initiator. Regions with a high frequency of development and imitation activities are characterised by dense networks for the inflow of ideas, new products and persons embodying specific competences (compare Johansson and Karlsson, 1987).

Once a specific design has been given to a product one can identify a date which signifies the vintage of the product. As long as the essential attributes remain unchanged, the vintage is considered to be the same. A large share of all new product variants which are created fail to enter and conquer a market. They disappear before they have been noticed. Among successful introductions of products we may distinguish between two basic cases: one with few and another with a large and expanding number of buyers. The latter constitutes the classical version of a life-cycle, along which the technology gradually matures as the result of a repeated introduction of new technical and commercial solutions. The price competition increases as the demand expands, the product becomes standardised and the production knowledge diffuses and becomes more public. When this process is prolonged in time, we can observe old product vintages.

Successful product development with only one or a few customers constitutes a second class, which is based on customisation such that every product is adapted to fit the preferences of the individual customer. The standard case is the delivery of equipment and systems to firms and organisations which are renewing their techniques of administration, production and distribution. In this type of situation, the attributes of every product and the pertinent delivery are adapted to match each customer's preferences as they are revealed in a willingness to pay for combinations of product attributes.

Let us now consider the vintage renewal of an individual production establishment, while at the same time the vintage of its product or product mix remains unchanged, which means that the vintage becomes older as time goes by. We start by introducing a strong simplification by examining a product which has unchanged attributes over its life cycle. By μ we denote the vintage of the product, and by τ we signify the technical vintage of the associated production process. Suppose that a product is introduced at $t = \mu$. If the technique is renewed every year, the age of the product at time t will be $t + \tau - \mu = \tau$.

The profit per unit output of a given technique τ can be written as (Johansson, 1987):

$$\pi(\tau) = p - g(\tau)\bar{c} \tag{1}$$

where p signifies the output price and $\bar{c}$ denotes a given cost level so that $g(\tau)\bar{c}$ is the unit production cost. The modernity of a process is described by increasing values of $\tau = 1, 2, \cdots$, when we use a discrete time scale. The different cost levels of various vintages are expressed by the function g such that $g(\tau) > g(\tau + v)$ for each $v > 0$. A basic assumption in product cycle theory is that the best practice technique is gradually improved over time. We assume that this process can be described as follows:

$$g(\tau) = \bar{g} + \tilde{g}\exp\{-\lambda\tau\} \tag{2}$$

where the oldest vintage is denoted by $\tau = 0$, and where λ is the rate at which the best practice technique is being improved. The formulation assumes that the technical knowledge develops over time, and that there is a lowest cost level below which the technique cannot be improved.

We assume that the price level $p(t)$ is reduced over time, as a consequence of increased price competition as time goes by and new technical vintages enter into the market. In the stylised model described by (1), the above development path is reflected by a falling price value so that $p(t) > p(t+v)$ for each t and all $v > 0$. This means that

$$\pi(\tau, t) = p(t) - g(\tau)\bar{c}$$

is being reduced as t grows. At time t, vintage τ has become economically obsolete if $\pi(\tau, t) \leq 0$. If input prices, including wages, increase over time, the cost factor $\bar{c}$ will have to be modelled as an increasing function of time, $\bar{c}(t)$, such that $d\bar{c}(t)/dt > 0$. For each given technical vintage, τ and θ, the economic ageing follows the type of path described in Figure 11.2. There we use the variable $\bar{c}^{\tau}(t) = g(\tau)\bar{c}(t)$.

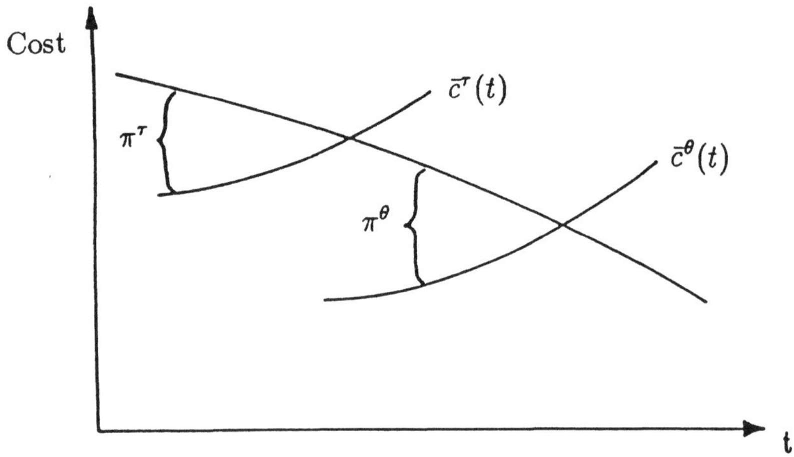

Figure 11.2 : Economics ageing of two successive production vintages τ and θ

11.2.2 Technological and Geographical Time Arrows

In the preceding Section we introduced a time arrow by assuming that $g(\tau)$ is a falling function of τ. However, we may still model a firm's accessibility to the knowledge as dependent on its *R&D* effort, the competence of its staff and the efficiency of its knowledge diffusion networks - which may be influenced by features of the region in which the firm/establishment is located.

Vast empirical studies show that $g(\tau)$ decreases for each new τ as the result of strict standardisation of the output and automation of the production, as well as the increasing scale of the production and distribution system (Førsund and Hjalmarsson, 1987; Wibe, 1980). The spatial consequence of this is that new vintages will be more sensitive to the cost of land and standard transport conditions. The spatial element can be brought into the model by applying a locational specification of the cost function in (1)-(2) such that

$$c^{\tau}(r) = p_h(r)h^{\tau} + p_b(r)b^{\tau} \tag{3}$$

where h and b are two categories of inputs and where $p_h(r)$ and $p_b(r)$ denote the associated prices in location r. This is a useful specification, since prices of input factors can be ordered in two groups - those which are relatively higher in locations with developed urban infrastructure attributes and those which are higher in more peripheral locations.

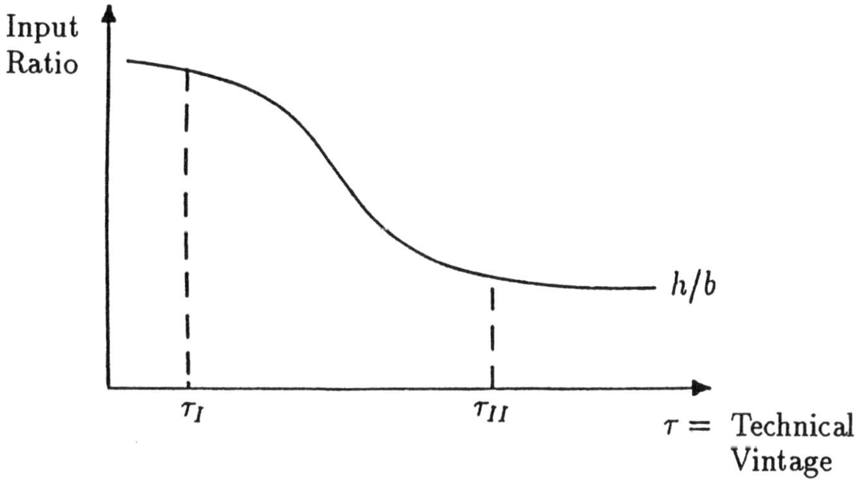

Figure 11.3 : Changing composition of input coefficients as the best practice vintage develops

The general idea behind the specification in (3) is to distinguish between the b-input and h-input resources. The latter are used for (i) knowledge acquisition, (ii) information about the preferences of potential and established customers, (iii) $R\&D$ and market development. The b-input resources are basic for automation, standardisation and large-scale solutions. With this classification of every production technique we can make the earlier assumption about a technological time arrow more precise. Thus, on the basis of assumptions from the product cycle theory we specify the following time arrow assumption:

> The improvement of techniques for producing a given product vintage follows a systematic path, such that the h-resources become successively less important over time, as new process vintages are developed. Hence, $h^\tau/b^\tau > h^{\tau+v}/b^{\tau+v}$ for each τ and all $v > 0$.

The above assumption about a technological time arrow is illustrated in Figure 11.3, which shows how the h/b-ratio is lowered as the technique-vintage develops from τ_I to τ_{II}.

We have already discussed our next assumption. It says that the urban infrastructure in central nodes differs from the built environment in other locations in such a way that the following price conditions prevail:

$$p_h(u)/p_h(r) < p_b(u)/p_b(r) \tag{5}$$

where u denotes a location in an infrastructure-rich urban region or metropolis, and where r signifies a more peripheral location, e.g. an export node. We may think of a line economy with a location variable ρ, such that $0 \leq \rho \leq u$ represents a location in urban infrastructure, while large ρ-values refer to decreasing centrality. When $\rho > r$ we refer to an economic periphery.

Consider now the assumptions in (4) and (5). Assume that actors choose locations which at each point in time minimize the location-specific cost function in (3). It is then shown in Andersson and Johansson (1984) how the location advantage is switched from location u to r as the best practice τ-index increases from $\tau = I$ to $\tau = II$. Such a switch is characterised by

$$\begin{aligned} p_h(u)h^I + p_b(u)b^I &< p_h(r)h^I + p_b(r)b^I \\ p_h(u)h^{II} + p_b(u)b^{II} &> p_h(r)h^{II} + p_b(r)b^{II}. \end{aligned} \tag{6}$$

The same type of switch is also demonstrated in Johansson and Westin (1987) by means of an analysis of the stationary points of all optimal locations for different τ-values.

$$H = \{(\rho, \tau) : \partial c^\tau(\rho)/\partial\rho = 0\} \tag{7}$$

where ρ denotes a location and H is the set of (ρ, τ)-combinations which minimise the cost function in (3). Differentiating the cost function a second time with respect to ρ we obtain

$$c^{\tau''}(\rho) = p_{\bar{h}}(\rho)h^\tau + p_{\bar{b}}(\rho)b^\tau. \tag{8}$$

Under general conditions one can show that there is a sequence of $\tau's$, where $\tau \rightarrow \tau_*$, such that for a given technical solution (h^{τ_*}, b^{τ_*}) the value of (8) is zero. That represents a bifurcation point at which the location advantage shifts from $\rho = u$ to $\rho = r$. This is illustrated in Table 11.1 which shows that the most advantageous location remains within the urban region for $\tau < \tau_*$. When technique τ_* becomes available the location advantage jumps from locations $\rho \leq u$ to locations $\rho > r$.[1]

The type of relocation process that we have described for a given product vintage illustrates that activities move out of nodes which have a rich urban infrastructure as new production techniques are developed and become available over time.

[1] Analytical details are provided in Johansson and Westin (1987). See also Garcia and Zangwill (1981).

Table 11.1 : A catastrophe shift in location advantage

$\rho < u$	$u < \rho < r$	$\rho > r$
$P\ddot{h} \gtrless 0$ $P\ddot{b} > 0$ $c^{\tau *''} > 0$	$P\ddot{h} < 0$ $P\ddot{b} > 0$ $c^{\tau *''} < 0$	$P\ddot{h} < 0$ $P\ddot{b} > 0$ $c^{\tau *''} > 0$
Location in a central urban node	No admissible solution in the distance interval (u, r)	Location in an economic periphery

$\star$ Analytical details are provided in Johansson and Westin (1987). See also Garcia and Zangwill (1981).

In this way space will be free for new activities in the urban region. In the next Section it is argued that this outmigration of maturing activities is a prerequisite for a repeated introduction of new activities based on new product vintages, where product signifies both goods and services. When the demand for additional space in an infrastructure-rich urban region becomes strong, the price variable $p_b(u)$ will increase further in comparison with $p_b(r)$. As a consequence the outmigration speed is increased, which will stimulate a faster entry of activities which can benefit enough from the u-location, and hence pay the price $p_b(u)$.

11.3 VINTAGE DYNAMICS OF URBAN INFRASTRUCTURE

11.3.1 Network Properties and Location Attributes

The analysis in the preceding subsection is focussed on the relocation from central urban nodes in which the pertinent young product vintages are initiated. In the maturing stage of the same vintage, the land price differential between a central node and the optional peripheral locations is the major force which influences the removal of the econômic activity from the central location with its high land values and rent levels. The importance of land price differentials may be illustrated by a model of von Thünen type. In such a model the price-bid functions of different activities vary with the distance from central and other concentrations of infrastructure-rich urban zones. In the 'post-industrial' epoch during the end of the 20th century we find that the steepest parts of the bid-price functions are related to negotiations and such development activities which mainly use the synergistic processes of human interaction. This includes the formation of commercial strategies and business alliances, the combination of design resources, the importation of technology and technicalequipment as well as scanning the world economy with regard to (i) the preferences of relevant customer groups, (ii) the emergence of new products, and (iii) the introduction of new technical solutions in general.

A new element in urban/metropolitan infrastructure is a growing importance of accessible corridors as location options which are alternatives to central core sites. The properties of the associated networks (which create attractivity) boil down to accessibility attributes of the person transportation systems. Those systems refer to commuting as well as person interaction as a part of economic activities. How can attractivity and high price-bids be explained against this background? As a general assumption, we suggest that the following features are basic attractor attributes:

(i) The possibility of adjusting individual time-tables for the coordination of meetings between persons

(ii) The existence of fast person transportation and accessibility in general

(iii) An integrated system of zones in which facilities, buildings and premises can be used for many types of person-interaction activities and by a broad scope of sectors.

From the conditions outlined in (i)-(iii) we can draw certain definite conclusions. The conditions describe basic attributes that make a metropolitan region and, in particular, urban zones of infrastructure attractive. This infrastructure may easily lose its productive value in a world where new product vintages continue to evolve. The dynamic mechanisms of the urban region can ascertain the usefulness of the infrastructure by continuously stimulating the exit of maturing and obsolete activities to other locations, while simultaneously facilitating the entry of new activities into the same urban environment (Figure 11.4). As we have explained earlier, price-bid behaviour is an important element of this self-organising evolution of a central urban node.

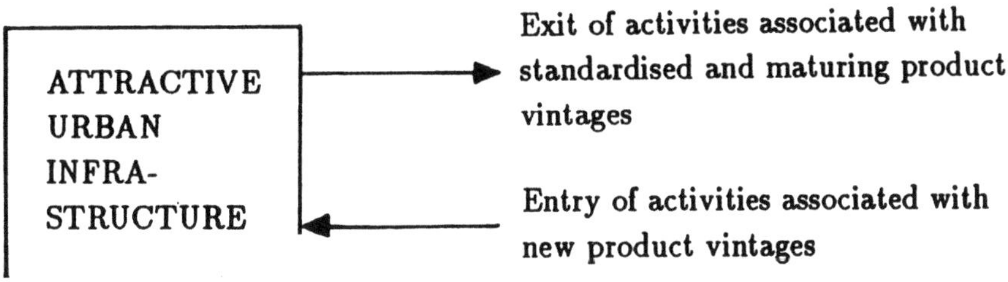

Figure 11.4 : Self-organised renewal of a vital urban environment

11.3.2 Exit and Entry Processes in Urban Infrastructure

The analysis in the preceding Section may be highlighted by a simple model of the exit and entry of space-using activities. We consider a gradual shift in an urban zone from old activities denoted by I and new, denoted by II. The space of the entire zone is signified by S. The amount of the two categories of activities is denoted by x_I and x_{II}. Now, introduce coefficients a_I and a_{II} which specify the amount of the activity space in the zone which is required for one unit of activity I and II, respectively. Suppose that the entry intensity is a function of the already established activity x_{II} and an attraction factor y so that we can specify $\dot{x}_{II} = dx_{II}/dt$ as follows:

$$\dot{x}_{II} = f(y, x_{II})[S - a_I x_I - a_{II} x_{II}]. \tag{9}$$

It is evident from equation (9) that if x_I is not being reduced or, if the pertinent process is very slow, then the entry will be delayed. When things work smoothly, $\dot{x}_{II} > 0$ and $\dot{x}_I < 0$ will develop along two symmetric sigmoid paths - one rising and one falling. In a long term perspective such gradual shifts can be interpreted as a sequence of economic episodes of a given urban zone or system of zones (Johansson and Nijkamp, 1987).

Typical phenomena that may disturb or delay the process in (9) can be identified as follows:

$$\dot{x}_{II} = f(y, x_{II})[S - a_I x_I(t-\omega) - a_{II} x_{II}(t) - a_0 \hat{x}_I(t)] \tag{10}$$

where the space $a_I x_I(t-\omega)$ becomes available for new activities with a time delay of ω. Moreover, if $\hat{x}_I(t)$ denotes the total amount of activity

I that has been removed since the decline of x_I started, then $a_0\hat{x}_I(t)$ represents the share of the pertinent land and the expanse of premises which are held back in a process of land-value speculation. It is obvious with this transparent picture of the evolutionary process that if the speed of entry is slowed down enough, the technology associated with episode II may become obsolete before it has had any chance to settle down in the urban environment we are discussing.

The replacement of I-activities by II-activities refers to a recurrent product cycle substitution process, in which standardised price-competing activities move out and non-standardised young product vintages move in and enter a phase of product competition. These new activities can enter more easily when the buildings and the pertinent premises function as infrastructure. This depends on the degree to which the buildings and facilities have been designed for general rather than special purposes. This has important dynamic implications. When activities move into stages of price competition, the gross profit is often reduced to a level which is too small to cover the rental costs in attractive and contact intensive zones of a central urban node. When such an activity leaves its building, other activities can move in without requiring much modification. Hence, this possibility of using the building capital in a sequence means that it does not represent sunk costs; instead it is infrastructure. This is furthermore stressed by the fact that the value of urban capital is associated with accessibility properties. Positive externalities are created in the form of contact potentials.

11.3.3 Renewal Cycles of Urban Infrastructure

Urban buildings and built environments constitute infrastructure when they satisfy certain generality conditions and have a flexible design. When this is the case, a building will during its life time have housed a sequence of different activities. Each individual activity may relocate or disappear depending on the vintage of its products and their associated economic age. This type of flexibility is a precondition for the existence and feasibility of metropolitan and urban regions in general. However, in a longer perspective we find that infrastructure components also embody technological features designed for specific needs within the urban network of buildings and interaction links. The pertinent technical solutions may also have a vintage property such that obsolescence and demolition may be analysed as ageing phenomena of successively outdated vintages. Section 11.4 presents empirical observations which show that urban zones exhibit vintage properties of a similar kind to those we find for production-specific capital, although the ageing of infrastructure is a very slow process.

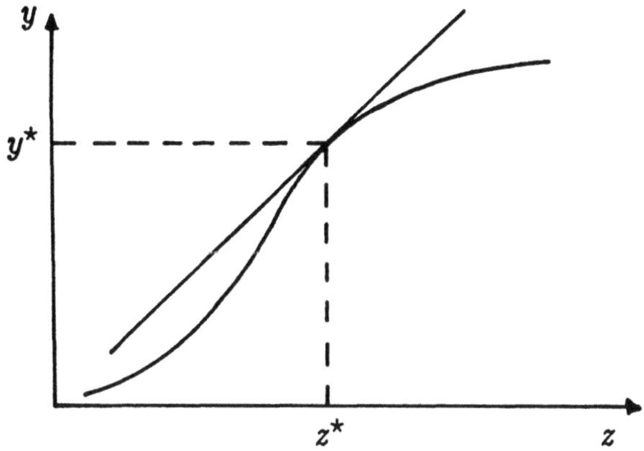

Figure 11.5 : Technological relation between production and infrastructure

In the sequel a metaphor is used to illustrate the above phenomena which comprise a dynamic interaction between the infrastructure of an urban zone and its production activities. The metaphor describes how the capacity and quality of a zone is developed gradually, how after a long period there is a sudden response with a fast economic expansion in the zone, and how the economic peak is followed by a very slow process of economic ageing. The metaphor also explains why, and how the process may be repeated. To start the story, let Z denote the amount of infrastructure and Y the volume of economic activities. Assume that the technologically optimal relation (in the sense of a production function) between the production Y and the infrastructure Z has the form illustrated in Figure 11.5, such that there is an initial phase of increasing returns to infrastructure scale, followed by a phase of decreasing returns. Formally we write

$$\begin{aligned} Y &= G(Z) \\ \hat{Z} &= H(Y) \end{aligned} \tag{11}$$

where H is the inverse of G and where $\hat{Z}$ represents the infrastructure volume which matches Y from a technological point of view. It is obvious from the Figure that the mapping H in (11) can be approximated as follows (compare Puu, 1984):

$$\hat{Z} = \lambda Y^3 - hY^2 + \sigma Y. \tag{12}$$

Let Z^* denote the point at which decreasing returns to infrastructure increments start, and let Y^* denote the associated level of economic activity as given by (11). Next, introduce

$$z = Z - Z^*$$
$$y = Y - Y^*. \tag{13}$$

Suppose that the infrastructure developer(s) in an urban region strives to allocate investment resources efficiently between zones, and to develop new zones when existing ones have reached their respective Z^*-levels. Such an inter-zonal priority principle may be reflected by the specification in formula (14). The specification assumes that the infrastructure decison-makers adjust their investments as follows

$$\dot{z} = -\alpha(Y - Y^*) = -\alpha y \tag{14}$$

where $Z^* = \beta Y^*$, as may be derived from a numerical specification of (11). A small value of α ascertains that the infrastructure adjusts slowly. Moreover, let

$$\hat{z} = \hat{Z} - Z^* \tag{15}$$

and assume that the economic activity adjusts in the following way:

$$\dot{Y} = \dot{y} = \theta(z - \hat{z}) \tag{16}$$

where $\theta = 1/\alpha$, and where $(z - \hat{z}) = (Z - \hat{Z})$. The self-organised adjustment in (16) may be explained as follows. When the infrastructure capacity is larger than what corresponds to the current economic activity, then there is an inflow of new and an expansion of already established activities. On the other hand, when the infrastructure is underbalanced, then the response is a reduction of the activity level in the zone.

Observe next that $\hat{Z} = H(Y)$ and that $Z^* = \beta Y^*$. Hence, we can find a function F such that $\hat{z} = F(y)$. As a consequence (16) can be rewritten as

$$\dot{y} = \theta(z - F(y)). \tag{17}$$

Observe that $z - \hat{z}$ in (16) can be written as $Z - Z^* - (Z - \hat{Z}^*)$, which means that we can use (12) to show that

$$F(y) = \lambda Y^3 - hY^2 + \sigma Y - \beta Y^*. \tag{18}$$

From the construction of Y^* we also know that at the point (Z^*, Y^*) the second-order derivative is zero so that $\partial^2 \hat{Z} / \partial Y^2 = 6\lambda Y^* - 2h = 0$, which means that $h = 3\lambda Y^*$. Moreover, at the same point we also have that $F(y) = 0$, which yields

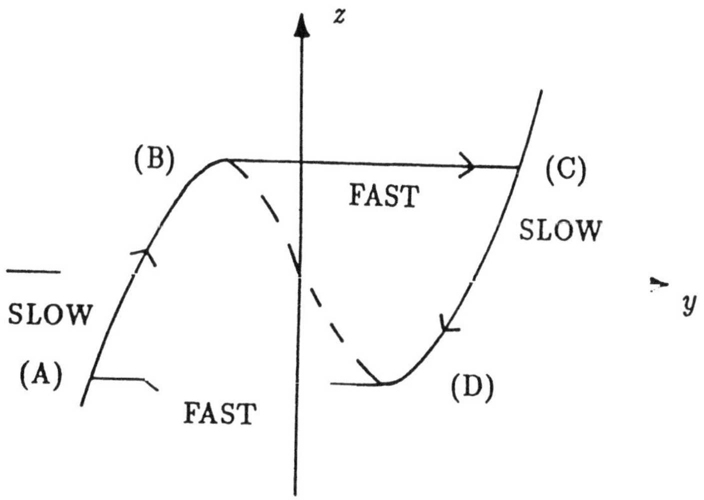

Figure 11.6 : Relaxation oscillations of infrastructure and production

$$(\beta - \sigma) = \lambda(Y^*)^2 - hY^*. \tag{19}$$

With these results given, we make the folowing substitution of variables:

$$\begin{aligned} h &= 3\lambda Y^* \\ \sigma &= hY^* - \mu \\ \beta &= \lambda(Y^*)^2 - \mu. \end{aligned} \tag{20}$$

Introducing the variables in (20) into (18) yields

$$F(y) = \mu y + \lambda y^3. \tag{21}$$

Observe now that the slow change of the infrastructure in urban zones corresponds to a small value of the parameter α in (14), and hence to a high value of the parameter θ in (17). Thus, y adjusts faster to deviations between z and $F(y)$ than z adjusts to the development of y. The equilibria of y are given by the equation $z = F(y)$ which leads to a cusp catastrophe, since on the pertinent time scale we treat z as a parameter. Next, we recognize the surface $z = F(y)$ as an equilibrium manifold in Figure 11.6. Off this set, the fast equation ensures that the trajectôries are nearly parallel to the y-axis. On the surface $\dot{y}$ vanishes so the system is governed entirely by the slow equation. With μ and λ positive, the (z, y)-pair moves autonomously around the orbit described in the Figure with sudden jumps at the relaxation points

(B) and (D). Hence, the system cycles around a closed loop in an alternating fast-slow, fast-slow periodic orbit. The time scales are of order α and θ for the fast and slow segments of the orbit, respectively (compare Lakin and Sanchez, 1982).

The metaphor provided in Figure 11.6 can be understood as follows. Let us start in point (A). This corresponds to an entirely new urban zone for economic activities - or an urban zone which is about to be completely renovated. The production level is far below its balanced value Y^*. Moreover, there is a lack of appropriate infrastructure, since Z is smaller than Z^*. As a consequence, efforts are made to improve the infrastructure. In a slow tempo the system then moves from (A) towards (B). As a part of this movement, the economic activity adjusts through a moderated but steady growth. Existing establishments extend their production volume and a small number of firms may move into the zone. At point (B) we have a situation with an oversized infrastructure which triggers a fast expansion of the economic activity. Soon congestion and other tensions become severe. All large investments in infrastructure have been stopped before point (C) has been reached, and now a slow deterioration period starts and establishments move slowly out of the zone, but still at a faster rate than that of the inflow of new production units. At point (D) the zone has become economically obsolete. The zone may now rather quickly become more or less empty of economic activities. In the model, the process may start all over again with renewal activities adapted to new vintages of economic life.

Which are the fundamental assumptions that generate the cyclic evolution pattern? Two assumptions are critical. The first is described in Figure 11.5 and Equation (12). It states that when the infrastructure size is small there are increasing returns to increments of the infrastructure. As the infrastructure size is augmented, a level Z^* is reached above which decreasing returns prevail. This should be considered a realistic assumption.

The second basic assumption is formulated in (14). It may be explained as a rule for infrastructure investors who have options to invest in many zones. In such a framework the target to find the balancing pair (Z^*, Y^*) of infrastructure and production in every zone, represents an attempt to distribute investment resources efficiently between zones. Equation (14) then says:

(i) If the economic activity increases beyond the target equilibrium Y^* for the zone, then the infrastructure capital should be allowed to decrease slowly.

(ii) When the economic activity in a zone is below the level Y^*, the infrastructure investments should continue.

Table 11.2 : Sales of information technology equipment by technology trade firms in the Stockholm region, 1986

PRODUCT VINTAGE	DELIVERY PATTERN IN PERCENT Stockholm region	Rest of Sweden	Exports
Less than 2 years old	40	15	10
2-5 years	35	23	20
More than 5 years old	25	62	70
	100	100	100

Source: Anderstig, Hårsman and Odén

Relying on arguments put forward earlier in this Chapter, we argue that this investment rule captures an important response pattern of infrastructure decision-makers (see also Snickars, 1984; Sundberg and Carlen, 1989; Peterson, 1990).

The response of the economic system to infrastructure supply, as given by (16), is straightforward. When the infrastructure in a zone is oversized vis-a-vis the level of economic activity, the economic activity continues to expand. When infrastructure is underbalanced, the opposite response is initiated. We may also make clear that the model as a whole tries to illuminate the intrazonal mechanisms. Hence, the interzonal conditions have to be surpressed and represented indirectly by conditions such as (14).

11.4 EMPIRICAL ILLUSTRATIONS

11.4.1 Exit, Entry and Local Networks

The product cycle mechanisms outlined in Section 11.2 were described for a given product vintage. At the same time we have presented a theoretical framework which explains why one should expect a frequent introduction of new product vintages in central urban nodes with a rich infrastructure for interaction between persons and for the establishment of novel activities. In Sweden the capital region (Stockholm)

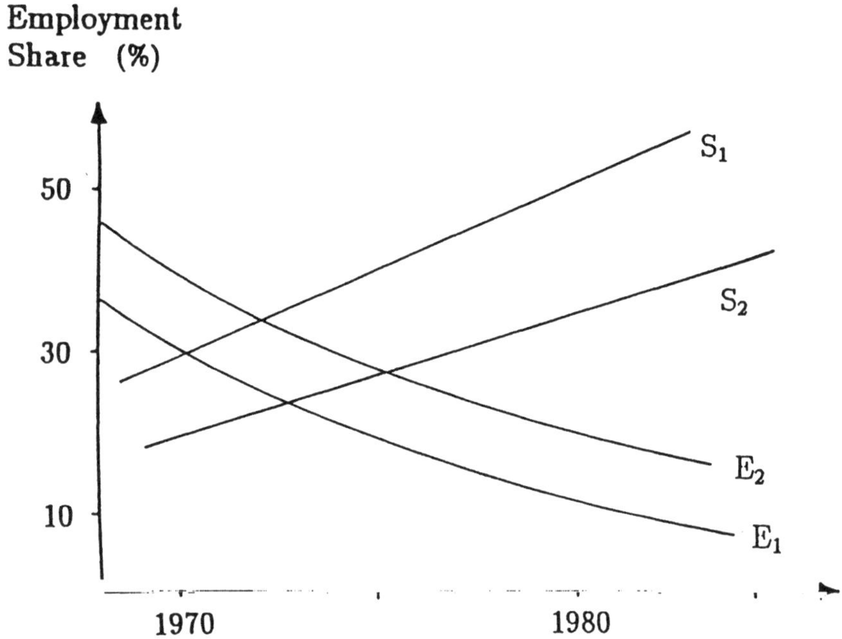

Figure 11.7 : Introduction of new and exit of old technical vintages in the manufacturing industry of Sweden. (Industrial Statistics of Sweden)

has been the main supplier of the above type of infrastructure in combination with international accessibility and a dense import network. We illustrate this in Table 11.2, which describes the market orientation of those technology trade firms which in 1986 were located in the Stockholm region and specialised in sales of information technology equipment. Around 60 percent of their sales were directed to Swedish regions outside Stockholm. However, when one separates different product vintages another picture emerges. For deliverers inside the Stockholm region, the share of products younger than 2 years was 40 percent. With regard to the rest of the country and the export market, the corresponding figures were 15 and 10 percent, respectively. Hence, the sales of the young product vintages are characterised by a concentration on local networks.

We have identified the Stockholm region as a central node in the Swedish multiregional economy. In such a node we expect a higher than average frequency of initiatives as regards product renewal and industrial reorganisation. We also expect the cost of premises and

Table 11.3 : Industrial zones in urban regions. Sweden 1975

Location	Zones established before 1920, %	Zones established after 1950, %
Central	66	6
Intermediary	27	22
Peripheral	7	72
	100	100

Source: Johansson and Strömqvist (1979)

other space to be higher than the regional average. Figure 11.7 illustrates the validity of these hypotheses. The Figure refers to the manufacturing industry as a whole. It describes the development of two variables S_1 and S_2, where 1 refers to Stockholm and 2 to the country as a whole. The S-variables denote the share of the labour force working in establishments with a productivity exceeding (at fixed prices) a given and high value μ^*. With this explanation it is obvious from the Figure that the Stockholm region behaves as expected; it has a faster rate of the introduction of new activities with a higher level of productivity than the national average.

In subsections 11.3.1 and 11.3.2 we emphasise that a fast expansion of novel activities, such as the development of S_1, usually requires a similarly fast outmigration of mature and obsolete production. To examine this we have also traced the share, E, of persons employed in manufacturing establishments with a low productivity μ^0, such that $\mu^0/\mu^* = 0.6$. In Figure 11.7 the Stockholm curve E_1 shows that establishments with a low level of productivity disappear earlier than in the country as a whole, which is represented by curve E_2.

An interpretation of Figure 11.7 says that the Stockholm region has managed to remain vital and to utilise its infrastructure by means of self-organising adjustments of the form described in Figure 11.4.

11.4.2 Obsolescence of Urban Zones

In an economy with technological development, new products (goods and services) and production techniques will be introduced and will phase out the old ones. The rate of economic change will be faster

Table 11.4 : Characteristics of urban zones in Sweden, 1975

Zonal attributes	Old zones* percent	Young zones** percent
Zones arranged in dense clusters	64	6
Located at harbour facilities	50	12
Concentration of unprofitable establishments	46	23
Zones in urban nodes with a contracting industry	40	22

Source: Johansson and Strömqvist (1979)

in environments in which the infrastructure is gradually adjusted and reshaped in order to accommodate the new activities. Even for general infrastructure there is a need for renewal as described in subsection 11.3.3.

Without change, the infrastructure will obstruct certain new activities from settling in a given urban node. In a similar way obsolete zones in an urban region will gradually be abandoned in favour of zones with suitable infrastructure. Illustrations of this phenomena can be collected from a comprehensive study of all urban regions in Sweden, such that more than one industrial zone could be identified in each urban region (Johansson and Stromqvist, 1979). The observation year is 1975; 24 urban regions hosting 280 zones were examined. The study shows that in older zones the infrastructure is gradually becoming unutilised, and the remaining production is characterised by economic obsolescence. Table 11.3 illustrates that the vintage structure of the urban regions are arranged as expanding circles according to age.

Table 11.4 illustrates that in the study the old zones are arranged in dense clusters. A large share of them are located near harbour facilities, a feature which has remained strongly correlated with economic obsolescence throughout the European urban regions during the period 1970-1990 (SOU 1990:34). Moreover, the share of economic obsolescence is much higher in the older than the younger zones.

11.4.3 Final Comments

This Chapter outlines a framework for analysing urban infrastructure, its network attributes and pertinent features with regard to logistical arrangements for person interaction. The efficiency of urban logistical arrangements has been discussed in terms of the possibility of fast adjustments of individuals' timetables for interaction. In order to further investigate the dynamics of economic innovations and infrastructure renewal, urban regions and their subsystems should be examined with regard to (compare Johansson and Nijkamp, 1987; Batten, 1989):

(i) The existence of threshold values related to endowments of infrastructure, labour force competence and *R&D* resources.

(ii) The presence of bottlenecks, revealed as congestion, price inflation etc.

(iii) Synergistic reinforcement when subsystems interact.

REFERENCES

Andersson, Å.E. and B. Johansson, 1984, "Knowledge Intensity and Product Cycles in Metropolitan Regions", WP-84-13, IIASA, Laxenburg, Austria.

Anderstig, C. and B. Hårsman, 1986, "On occupation structure and location patterns in the Stockholm region", *Regional Science and Urban Economics*, 16:97-122.

Anderstig, C., Hårsman, B. and N. Odén, 1986, "New Industrial Structures", Rapport 1986:4, Regionplanekontoret, Stockholms Läns Landsting (in Swedish).

Batten, D.F., 1989, "Infrastructure as a Network System - Mera Revisited", CWP-1989:22, CERUM, Umeå University.

Currea, D.P. and K.R. Polenske, 1985, "Planning for Public Infrastructure Investment", in F. Snickars, B. Johansson, and T.R. Lakshmanan (eds.), *Economic Faces of the Building Sector*, Document D20:1985, Swedish Council for Building Research, Stockholm.

Førsund, F. and L. Hjamarsson, 1987, *Analysies of Industrial Structure: A Putty-Clay Approach*, The Industrial Institute for Economic and Social Research, Stockholm.

Garcia, C.B. and W.I. Zangwill, 1981, *Pathways to Solution, Fixed Points and Equilibria*, Prentice-Hall, New York.
Johansson, B., 1987, "Technological Vintages and Substitution Process" in D. Batten, J. Casti and B. Johansson (eds.), *Economic Evolution and Structural Adjustment*, Springer-Verlag, Berlin.
Johansson, B., 1989, "Metropolitan Nodes in the Innovation Networks of the Nordic Economies", in *The Long Term Futures of Regional Policy - A Nordic View*, Nord REFO / OECD, Borgå, 1989.
Johansson, B. and C. Karlsson, 1987, "Processes of Industrial Change: Scale, Location and Type of Job" in M. Fischer abd P. Nijkamp (eds.), *Regional Labour Markets*, North-Holland, Amsterdam, pp. 139-165.
Johansson, B. and P. Nijkamp, 1987, "Analysis of Episodes in Urban Event Histories", in L. van der Berg, L. Burns and L.H. Klaassen (eds.), *Spatial Cycles*, Gower, Aldershot, pp. 43-66.
Johansson, B. and F. Snickars, 1988, "Modelling the economic dynamics of knowledge-intensive metropolis", *Sistemi Urbani*, 63-91.
Johansson, B. and U. Strömqvist, 1979, "Industrial Zones in Swedish Urban Regions", Rapport R24:1979, Swedish Council for Building Research, Stockholm (in Swedish).
Johansson, B. and L. Westin, 1987, "Technical change, location and trade", *Papers of the Regional Science Association*, 62:13-25.
Lakin, D. and D.A. Sanchez, 1982, *Topics in Ordinary Differential Equations*, Dover Publications, New York.
Lakshmanan, T.R., 1989, "Infrastructure and Economic Transformation", in Å.E. Andersson, D.F. Batten, B. Johansson and P. Nijkamp (eds.), *Advances in Spatial Theory and Dynamics*, North-Holland, Amsterdam, pp. 241-262.
Mera, K., 1984, *Measuring Economic Contributions of Infrastructure in Cities of Developing Countries*, UUD-54, Water Supply and Urban Development Department, The World Bank, Washington D.C.
Okada, N. and K. Kobayashi, 1989, "The Region as a Creative Forum - A Conceptual Approach", CWP-1989:23, CERUM, Umeå University, Umeå.
Peterson, G.E., 1990, "Declining Infrastructure Investments", in *Urban Challenges*, SOU 1990:33, Allmanna Förlaget, Stockholm.
Puu, T., 1984, "A Class of Homothetic Production Functions Obeying the Frischian Regular Ultra Passum Law", Umeå Economic Studies No, 9, Umeå University, Umeå.
Puu, T., 1988, "Optimality versus Stability in Spatial Economic Pattern Formation", CWP-1988:29, CERUM, Umeå University, Umeå.
Snickars, F., 1989, "Effects of Infrastructure Provision on Urban Economic Development", CWP-1989:2, CERUM, Umeå University.

SOU, 1990, *Urban Regions in Europe*, SOU 1990:34, Allmänna Förlaget, Stockholm (in Swedish).
Sundberg, L. and G. Carlén, 1989, "Allocation mechanisms in public provision of transport and communication infrastructure", *The Annals of Regional Science*, 23:311-327.

CHAPTER 12

Information Technology and Urban Spatial Structure

Komei Sasaki

12.1 INTRODUCTION

The recent striking progress in information-communication technology has made it possible to store, process, and transmit a large amount of information in a short time at a cheap cost. In fact, many firms and households are equipped with the installations for information technology. This tendency is called '*Johoka*' (informationization) in Japan. Even parts of the public sector, such as local governments, have started to establish the infrastructure for telecommunications with the hope that it will promote the regional economy. Improvements in information technology will cause some changes in a city or region since the behaviour of agents operating and residing there, such as firms and households, will be affected. However, what will change to what extent is not unambiguous. Various statements have been made concerning the probable impacts of progress in information technology. Most of them, however, seem just conjecture, or wishful thinking which is not necessarily predicted in a scientific way. Thus, they may be 'unsubstantiated assertions'.[1]

The decentralization of firms in a city is predicted to be a typical feature of the information society. In particular, the routine-activity sector of a firm might tend to be separated from the head-office and located in a suburban area where land rents as well as wage rates are lower.[2] As far as households are concerned, a popular scenario is that

[1] Pressman (1985, p.357).

[2] For instance, Sakamura (1989) stresses that there is necessarily a shift from centralization to decentralization in a city as 'informationization' develops.

as information technology pervades them, people will work at home rather than at offices, using TV- or Tele-shopping instead of visiting stores. As a result, the distance travelled by city residents will become significantly shorter. However, whether or not firms become decentralized not only depends on the level of information technology, but on other factors such as output, employment, production technology, wage rates, and commuting costs. [3] It is therefore necessary to clarify the conditions under which the decentralization of firms will take place. Also, the following questions need to be answered. How is the location of residences affected and how is the land rent structure changed as firms are decentralized? Is the volume of city traffic increased or decreased as information technology pervades a city?[4]

The present Chapter intends to answer, to some extent, the above questions concerning the impacts on the urban structure of the pervasion of information technology, focusing on the location behaviour of firms, especially on the possibility of separate head-office and routine function locations. Within the framework of the non-monocentric model of Ogawa and Fujita (1980) we will demonstrate the possible patterns of urban configurations in equilibrium and examine the conditions under which each of them emerges.

In Section 12.2 the behaviour of a firm and a household is described. Section 12.3 will explain the bid-rent function, which is the central concept in this Chapter; it is also devoted to the preparatory analysis for classifying the equilibrium configuration patterns of a city. In Section 12.4, the conditions for each of the possible configuration patterns to emerge are fully investigated. Based on the analyses in Section 12.4, Section 12.5 intends to predict how the urban configuration is changed as information technology spreads throughout a city.

12.2 THE MODEL

12.2.1 Behaviour of a Firm

The activities of a firm can be, roughly, classified into two functions: administrative functions and routine-functions. The former includes decision making, planning, and transactions and negotiations with other firms and customers, while the latter is concerned with production, shipments, the management of inventories, bookkeeping, collecting information at the plant level and transmitting information

[3] Pressman (1985, p.360).

[4] Salomon (1986) emphasizes that the relation between tele-communication and transportation demand is not unidirectional.

to executives. We assume that these two activities can be separated physically as well as functionally, and whether or not they are located in the same place or in different places depends on the particular circumstances of a firm. The two functions are located in the same place or very close to each other when communication between them costs a lot due to the physical distance between them. On the other hand, when the communication cost between two functions is not large, and the routine-function requires a large amount of labour and extensive land, the routine-activity sector tends to be located in a surburban area while the head-office is located in the city centre.

Firms are involved in two different grades of communication trips. We specify these two grades as follows. High-grade communication trips originate at the head-office in order to transact and bargain with customers, and to make contact with other firms for collecting information about business. It is presumed that all these high-grade forms of communication are made and conducted on the 'trading floor' established at the city centre. The high-grade communication is carried out in face-to-face meetings since the information obtained by this form of communication is unstandardized and of a high quality. It must be emphasized that face-to-face contact for the high-grade forms of communication will not be replaced by telecommunications or the like, even when information technology becomes commonly available. The second type of communication is low-grade, for the collection and transmission of low-quality, standardized information. Trips for this grade of communication are generated in the routine-activity sector, and have two different destinations: one is the trading floor in the city centre and the other is the head-office of the firm. Employees of a routine-activity sector collect information about the movements of the demand for products, wage rates, land rents, and the prices of materials on the trading floor; they also report the data on current production, sales, and inventory stocks to the head-office and receive instructions about various plans from the executives. Since this low-grade communication is standardized, and a large amount of information is collected and transmitted, trips for this form of communication can be replaced by means of information technology.[5]

In this Chapter we assume a long narrow city where the distance in the breadthwise direction is negligible. Let us also assume that the trading floor has already been established at the centre of this long narrow plain (called location 0) and that any location in this city can be represented in terms of the distance from the centre. In this city, $2N_f$ firms with identical technologies and size are operating to produce

[5]Evans (1985, pp.50-64) discusses the relationship between the quality of information and the methods for transmitting information. The terminology of the 'trading floor' originated in his book.

a homogeneous product (or service), and $2N$ households with identical tastes reside and work for $2N_f$ firms. No generality is lost by treating only either half of a city on the assumption that an equilibrium urban configuration is symmetric around the centre. Thus, only the section to the right of the centre is treated in the subsequent analysis. The present Chapter focuses on the possible patterns of urban configurations in equilibrium (which are characterized by a spatial distribution of firms and households, the distance between the head-office and the routine-functions of a firm, the wage profile, and the land rent structure), and the conditions for each pattern to emerge. A more general framework where the location area of each agent is not pre-specified is required for this purpose. From this viewpoint the non-monocentric model developed by Ogawa and Fujita (1980) is appropriate.

Assume the head-office and routine-activity sectors of a firm are located, respectively, at x_0 and x_1 (in terms of distance from the centre). A firm's operation is based on a production function of a fixed-coefficient type. The head-office and the routine-function sector, respectively, need the capital service c_0 and c_1, labour l_0 and l_1, and land size h_0 and h_1 in order to produce and sell unit output. With respect to labour input, it is assumed more labour is employed in routine-functions, i.e., $l_1 > l_0$ (which seems realistic). As described above, in addition to the three inputs a firm needs to make a certain number of trips for communications of two different grades in order to realize the production and sale of unit output. A firm incurs the trip cost $g_0(S(x_0, 0))$ for high-grade communications, and $g_1(S(x_0, x_1), S(x_1, 0))$ for low-grade communications in producing and selling unit output, where $S(x_i, x_j)$ denotes the physical distance between locations x_i and x_j. It is assumed that $dg_0/dS > 0$, $d^2g_0/dS^2 > 0$, $\partial g_1/\partial S(x_0, x_1) > 0$, $\partial^2 g_1/\partial S^2(x_0, x_1) > 0$, $\partial g_1/\partial S(x_1, 0) > 0$, $\partial^2 g_1/\partial S^2(x_1, 0) > 0$, $\partial^2 g_1/\partial S(x_0, x_1)\partial S(x_1, 0) = 0$. From the definitions so far, the profit of a firm locating its head-office at x_0 and its routine-activity sector at x_1 is represented as follows:

$$\begin{aligned}\pi(x_0, x_1) \quad = \quad & Q(x_0, x_1)\{p - \rho(c_0 + c_1) - w(x_0)l_0 \\ & -w(x_1)l_1 - r(x_0)h_0 - r(x_1)h_1 - g_0(S(x_0, 0)) \\ & -g_1(S(x_1, x_0), S(x_1, 0))\}\end{aligned} \tag{1}$$

in which $Q(x_0, x_1)$ = the output, $w(x)$ = the wage rate at x, $r(x)$ = the land rent at x, p = the price of the product, ρ = the price of the capital service. In order to simplify the analysis it is assumed that the output level of a firm is the same regardless of its location and given a priori: i.e., $Q(x_0, x_1) = \bar{Q}$. Each firm chooses the locations of the head-office and routine-activity, $\{x_0, x_1\}$ so as to maximize the profit given $\{p, \rho, w(x), r(x)\}$. The profit attained at equilibrium is the same for each firm because of the homogeneity of firms: i.e., $\pi(x_0^*, x_1^*) = \pi^*$.

The wage rates and land rents are determined so as to ensure the same profit. As we described above, the pervasion of information technology will affect the cost function for the low-grade communication g_1. In order to explicitly analyze the impact of information technology, the communication cost functions need to be expressed in more concrete forms. They are specified in the following way:

$$g_0 = nS(x_0, 0)^2$$
$$g_1 = m\{S(x_1, x_0)^2 + S(x_1, 0)^2\} \tag{2}$$

where $S(x_i, x_j) = |x_i - x_j|$. In relation (2), $n \gg m$ since high-grade communications are performed by executives of a firm and hence the higher value of their time is reflected in the trip cost per distance.[6]

12.2.2 Behaviour of a Household

A household derives its utility from the consumption of composite goods and residential lot size: that is

$$u = u(a, q) \tag{3}$$

where a = a composite good, q = the lot size of a residence. The budget constraint of a household residing at x_r and commuting to x_w is represented by

$$w(x_w) = a + r(x_r)q + k|x_w - x_r| \tag{4}$$

where k denotes the commuting cost per unit distance and the price of the composite good is unity. Throughout this Chapter the residential lot size is assumed to be the same regardless of location and is given a priori for the simplification of analysis: i.e., $q = \bar{q}$. Each household chooses the locations of residence and workplace, $\{x_r, x_w\}$ so as to maximize its utility (and hence a) given $\{r(x), w(x), k\}$. Since households are homogeneous, the utility level in equilibrium is the same amongst all of them: i.e., $u(a^*, \bar{q}) = u^*$.

We have sketched the behaviour of two agents so far. We note that without some restrictive assumptions (in particular, $Q = \bar{Q}$ and $q = \bar{q}$) it is impossible to develop the analysis within a general framework. These assumptions are made in Ogawa & Fujita (1980) as well, and might be considered as some kind of cost for using a general model.

[6] In the present model all the residents in a city are assumed identical. A few executives working for head-offices, distinguished from the *ordinary* workers, are supposed to reside outside a city or inside the building of the head-office. The costs of face-to-face contacts made by executives might include the expenses for entertaining customers. Such expenses do not change with the distance of trips, and are assumed to be included in the capital cost ρC. Executive salaries are assumed to be included in n to account for the time cost of executive trips.

12.2.3 Commuting Patterns, Wage Profiles, and the Location of Firms

Before proceeding to the equilibrium urban configuration, the following three properties are demonstrated.

Property 1 (commuting patterns): *In the equilibrium land use pattern, no cross-commuting takes place where every household has a strict preference for its own job site.*

Cross-commuting is such that for the two households whose locations of residence and workplace are $\{x_r, x_w\}$ and $\{x'_r, x'_w\}$, respectively, both $x_r < x'_r$ and $x'_w < x_w$ hold. That a household residing at x_r has a strict preference for the job site x_w compared with x'_w implies that the net income obtained by working at x_w is strictly larger than that of x'_w. See Ogawa & Fujita (1980, p.462) for the proof of Property 1.

Property 2 (wage profiles):

$$\text{If } x_r < x'_r \leq x_w < x'_w \text{ , then } \frac{dw(z)}{dz} = k \text{ at each } z \in [x_w, x'_w] \text{ .}$$

$$\text{If } x_w < x'_w \leq x_r < x'_r \text{ , then } \frac{dw(z)}{dz} = -k \text{ at each } z \in [x_w, x'_w].$$

Property 2 implies that a wage function in equilibrium is linear within the business area. See Ogawa & Fujita (1980, pp.463-4) for the proof.

Before proceeding to Property 3, we will introduce some new terminologies. Suppose a firm locates its head-office at x_0. If it holds that $\pi(x_0, x_1) > \pi(x_0, x'_1)$, then we say that this firm has a strict preference for x_1 compared with x'_1 as the location of its routine-functions. Suppose two firms who locate their two functions at $\{x_0, x_1\}$ and $\{x'_0, x'_1\}$. If $x_0 < x'_0$ and $x'_1 < x_1$, then we call this state a cross-location of the routine-function.

Property 3: *No cross-location of the routine-function takes place in the equilibrium where each firm has a strict preference for the location of its routine-function compared with any other locations. (See Sasaki (1990) for the proof.)*

12.3 BID-RENT FUNCTION

It is assumed that all land in this city is owned by absentee landlords, and any parcel of land is rented to an agent offering the highest bid-rent for that lot. Therefore the market land rent of a particular lot in equilibrium is equal to the bid-rent offered by an agent locating there.

12.3.1 Bid-Rent of a Firm

Suppose a firm planning to locate its head-office at x_0. The bid-rent of the firm for a parcel of land at x_0 is defined as follows:

$$\begin{aligned}\Phi(x_0) &= \max \frac{1}{h_0}[p - \rho(c_0 + c_1) - w(x_0)l_0 - w(x_1)l_1 \\ &\quad - r(x_1)h_1 - n{x_0}^2 - m\{(x_1 - x_0)^2 + {x_1}^2\} - \tilde{\pi}]\end{aligned} \tag{5}$$

where $\tilde{\pi}$ is the profit per unit output obtained in equilibrium ($= \pi^*/\bar{Q}$). The necessary condition for (5) is:

$$\frac{d\Phi_0}{dx_1} = -\frac{1}{h_0}[w'(x_1)l_1 + r'(x_1)h_1 + 2m\{(x_1 - x_0) + x_1\}] = 0. \tag{6}$$

The gradient of a bid-rent function for the head-office is derived by Envelope theorem as

$$\frac{d\Phi_0}{dx_0} = -\frac{1}{h_0}[w'(x_0)l_0 + 2nx_0 - 2m(x_1 - x_0)]. \tag{7}$$

Taking into consideration that a change in the location of head-office x_0 causes a change in the location of routine-function x_1, we have (8) as the second derivative.

$$\frac{d^2\Phi_0}{dx_0^2} = -\frac{1}{h_0}[w''(x_0)l_0 + 2n - 2m\{\frac{dx_1}{dx_0} - 1\}]. \tag{8}$$

Meanwhile, the bid-rent of a firm locating its head-office at x_0 for a parcel of land at x_1 to locate its routine-function is defined as follows:

$$\begin{aligned}\Phi_1(x_1) &= \frac{1}{h_1}[p - \rho(c_0 + c_1) - w(x_0)l_0 - w(x_1)l_1 \\ &\quad -\Phi_0(x_0)h_0 - n{x_0}^2 - m\{(x_1 - x_0)^2 + {x_1}^2\} - \tilde{\pi}].\end{aligned} \tag{9}$$

$\Phi_1(x_1)$ must be equal to $r(x_1)$ when that firm actually locates its routine-function at x_1. Interpreting the slope of the bid-rent curve for the routine-function at x_1 is not straightforward since the location of the routine-function does not change freely but depends on the location of head-office x_0. Differentiating $\Phi_1(x_1)$ with respect to x_0, we obtain

$$\frac{d\Phi_1(x_1)}{dx_0} = \frac{\partial\Phi_1(x_1)}{\partial x_0} + \frac{\partial\Phi_1(x_1)}{\partial x_1}\frac{dx_1}{dx_0}. \tag{10}$$

From the relation in (7), the first term on the right-hand side in (10) is equal to zero. If

$$\frac{dx_1}{dx_0} = 1 \tag{11}$$

then it follows that $d\Phi_1/dx_0 = \partial\Phi_1/\partial x_1$: the slope of a bid-rent function for the routine-activity is equal to a variation in the bid-rent of the routine-function in response to a marginal change in the location of the head-office. In this Chapter, the location pattern specified by (11) is assumed: if the head-office of a firm becomes farther from the city centre than before (or from the head-office of other firms) by unit distance, then its routine-function becomes farther from the city centre than before (or from the routine-function offices of other firms) by unit distance. Under this assumption (8) is reduced to

$$\frac{d^2\Phi_0}{dx_0{}^2} = -\frac{1}{h_0}[w''(x_0)l_0 + 2n]$$

and it holds concerning the bid-rent function of routine-activity that

$$\frac{\partial\Phi_1(x_1)}{\partial x_1} = -\frac{1}{h_1}[w'(x_1)l_1 + 2m\{(x_1 - x_0) + x_1\}] \tag{12}$$

$$\frac{d(\frac{\partial\Phi_1(x)}{\partial x_1})}{dx_0} = -\frac{1}{h_1}[w''(x_1)l_1 + 2m]. \tag{13}$$

12.3.2 Bid-Rent of a Household

The bid-rent of a household planning to reside at x_r is defined as

$$\Psi(x_r) = \max_{x_w} \frac{w(x_w) - a(q, u^*) - k|x_w - x_r|}{q}. \tag{14}$$

The necessary condition for (14) is

$$\begin{aligned} w'(x_w) &= k \quad \text{when} \quad x_w \geq x_r \\ &= -k \quad \text{when} \quad x_w < x_r. \end{aligned} \tag{15}$$

By the Envelope theorem the gradient of a bid-rent function is derived as follows:

$$\begin{aligned} \frac{\partial\Psi(x_r)}{\partial x_r} &= \frac{k}{q} \quad \text{when} \quad x_w \geq x_r \\ &= \frac{-k}{q} \quad \text{when} \quad x_w < x_r. \end{aligned} \tag{16}$$

12.3.3 Preliminary Analysis

We have explained the basic framework. There are some alternative patterns of urban configurations, the configurations being characterized by the locations of firms' head-offices and routine-functions and households' residences. Before examining each of them we will present the following two properties.

> Property 4: *The routine-function of a firm is never located closer to the city centre than its head-office. (For a detailed proof, see Sasaki, 1990.)*

> Property 5: *If the head-office and the routine-activity sectors of each firm are located separately from each other, then the head-office of a firm and the routine-functions of other firms are never located at the same place. (For a detailed Proof, see Sasaki, 1990.)*

12.4 URBAN SPATIAL STRUCTURE IN EQUILIBRIUM

There are some alternative patterns of urban configurations to be realized in equilibrium. Following Ogawa & Fujita (1980), they are classified into two patterns: i.e., *exclusive land use patterns* (what is termed a 'connecting pattern', Ogawa & Fujita, 1980), and *mixed land use patterns*. In an exclusive land use pattern, any location in a city is used exclusively by either firms or households: there is no coexistence of firms and households at any location. On the other hand, in a mixed land use pattern, all of the lots or some of them are used for both business activities and residences; we call it *a completely mixed pattern* when all of the lots in a city are used in mixed ways, and an *incompletely mixed pattern* when only some of the lots are used in mixed ways. As long as the land rent for urban use is not below the agricultural land rent, there is no vacant land in a city in equilibrium. (See Ogawa & Fujita, 1978, pp.21-26.)

12.4.1 Exclusive Land Use Patterns

By virtue of Property 4, the routine-function of a firm is not placed at a location closer to the city centre than its head-office (i.e., $x_0 \leq x_1$). In this situation, Property 6 applies.

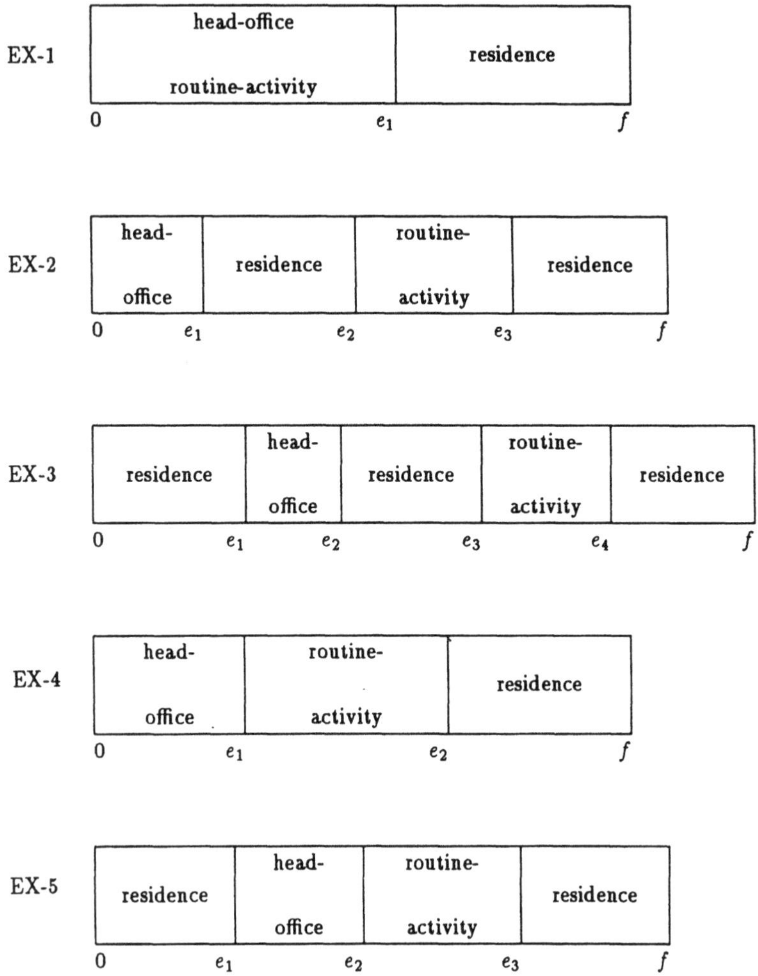

Figure 12.1 : Exclusive land use patterns

Property 6: *There is no outward commuting to the routine-function sectors.*

We see immediately from Property 6 that routine-function sectors are not located in the periphery of a city. From the arguments so far, there are five alternative configurations of equilibrium exclusive land use patterns, as shown in Figure 12.1.

We can derive the equilibrium conditions in (17) through (21) for five alternative land use patterns and depict the corresponding land rent structures in Figures 12.2 through 12.6. (For the detailed process of derivation, see Sasaki, 1990.) We note that $h_0 = h_1 = h$ needs to be assumed in any of EX-2 — 5 cases in order to ensure that the distance between the two functions is identical among firms.

$$\text{EX}-1 \qquad (m+n) \geq \frac{k(l_0+l_1)}{2N}[\frac{1}{q}+\frac{l_0+l_1}{h_0+h_1}] \tag{17}$$

$$\text{EX}-2 \qquad \min\{\frac{2n\bar{Q}N_fh-k(\frac{h}{q}+l_0)}{2\bar{Q}N_f(h+l_0q)}, \frac{k(\frac{h}{q}+l_1)}{4\bar{Q}N_f(h+l_0q)}\}$$

$$\geq m \geq \frac{k(\frac{h}{q}+l_1)}{2\bar{Q}N_f(3h+2l_0q)} \tag{18}$$

$$\text{EX}-3 \qquad \min\{\frac{2n\bar{Q}N_fh-k(\frac{h}{q}+l_0)+2(n+m)e_1}{2\bar{Q}N_f(h+ql_0)}$$

$$, \frac{k(\frac{h}{q}+l_1)+2me_1}{4\bar{Q}N_f(h+ql_0)}\} \geq m \geq \frac{k(\frac{h}{q}+l_1)+2me_1}{2\bar{Q}N_f(3h+2ql_0)} \tag{19}$$

$$\text{EX}-4 \qquad \frac{2n\bar{Q}N_fh+k(l_1-l_0)}{6\bar{Q}N_fh} \geq m \geq \frac{k(\frac{h}{q}+l_1)}{6\bar{Q}N_fh} \tag{20}$$

$$\text{EX}-5 \qquad \frac{2n\bar{Q}N_f(h+ql_0)+k(l_0+l_1)}{2\bar{Q}N_f(4h+ql_0)}$$

$$\geq m \geq \max\{\frac{2n\bar{Q}N_fql_0}{2\bar{Q}N_fh}, \frac{k(\frac{h}{q}+l_1)}{2\bar{Q}N_f(3h+ql_0)}.\} \tag{21}$$

In the figures, R_A represents the agricultural rent. It is proved that each configuration cannot be repeated to the right of f, and thus we have:

Property 7: *In an equilibrium of an exclusive land use pattern, only one district is formed for each head-office and routine-function office in the right (or left) half of a city. (For a detailed proof, see Sasaki, 1990.)*

12.4.2 Mixed Land Use Patterns

First we show the following property with respect to the mixed land use pattern.

Property 8: *Households residing in the mixed district where business activities and residences coexist are employed in their own residential locations (i.e., $x_r = x_w$). (For a detailed proof, see Sasaki, 1990.)*

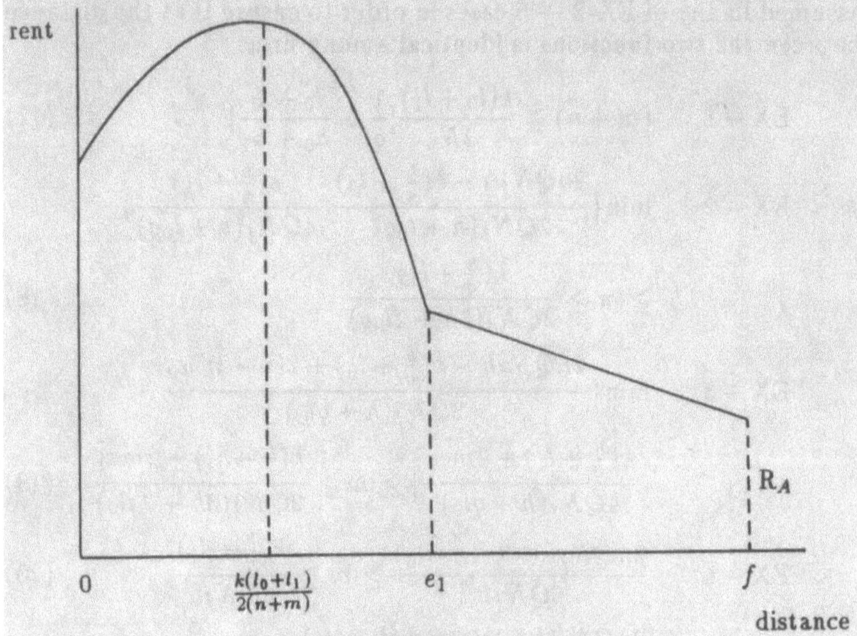

Figure 12.2 : Land rent structure of EX-1

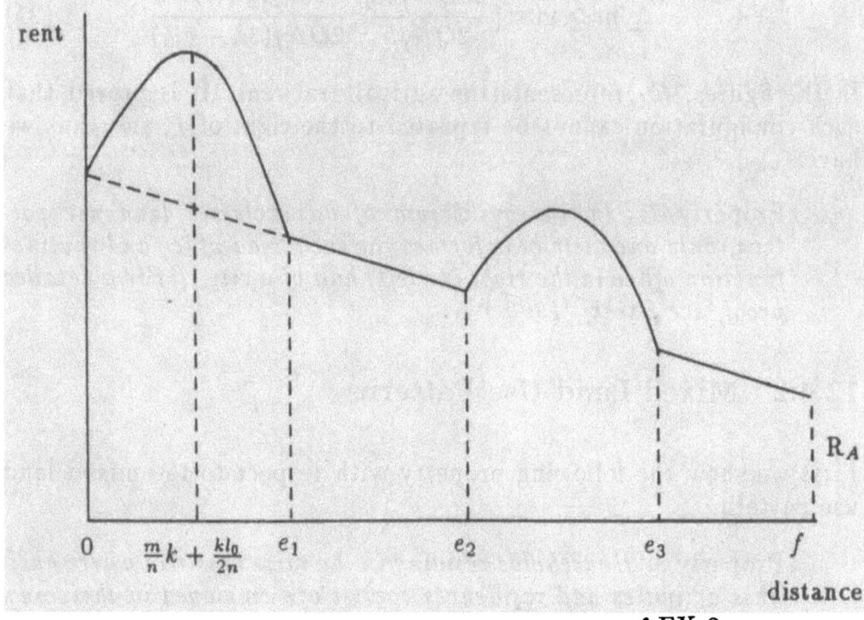

Figure 12.3 : Land rent structure of EX-2

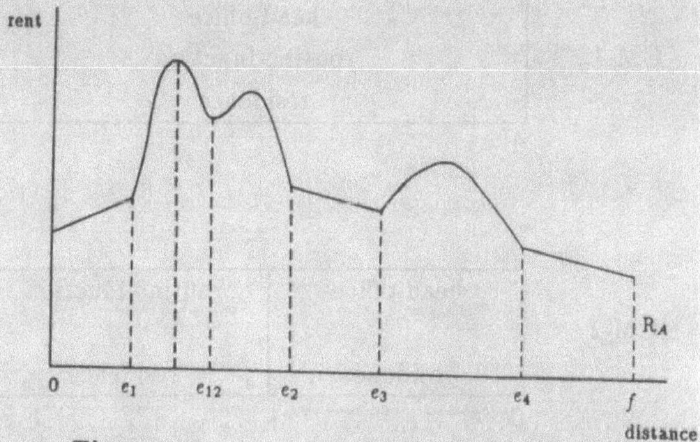

Figure 12.4 : Land rent structure of EX-3

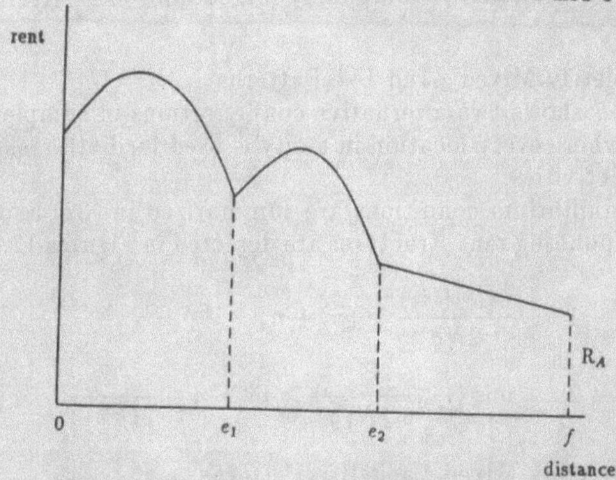

Figure 12.5 : Land rent structure of Ex-4

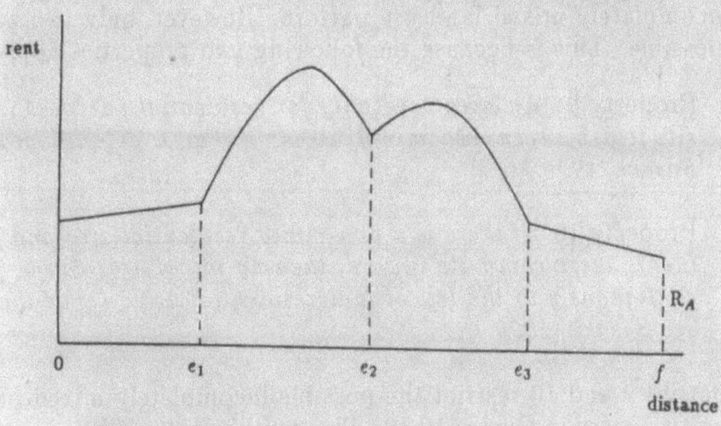

Figure 12.6 : Land rent structure of EX-5

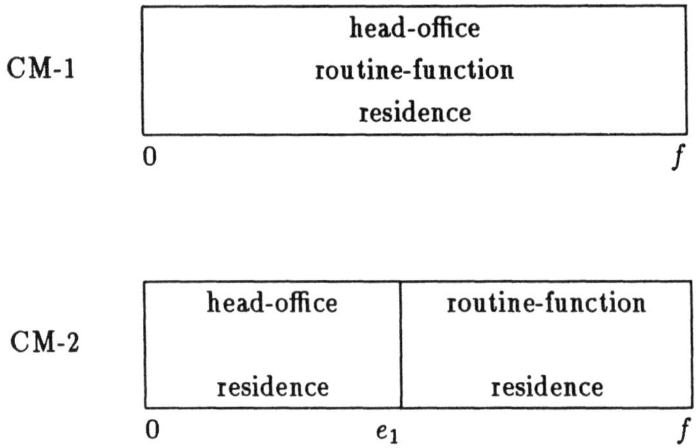

Figure 12.7 : **Completely mixed land use patterns**

(1) Completely Mixed Land Use Patterns
Figure 12.7 shows two alternative configurations of completely mixed land use where every location in a city is used for both residential and business activities.

The equilibrium conditions are summarized in (22) and (23), and the corresponding rent structures are depicted in Figures 12.8 and 12.9.

$$\text{CM} - 1 \qquad \frac{k(l_0 + l_1)}{2Nq} \geq (m + n) \tag{22}$$

$$\text{CM} - 2 \qquad \min\{\frac{n}{3}, \frac{k}{6\bar{Q}N_f q}\} \geq m \geq n - \frac{k}{2\bar{Q}N_f q}. \tag{23}$$

(2) Incompletely Mixed Land Use Patterns
There might seem to be many variations in the urban configuration of an incompletely mixed land use pattern. However, only two patterns are possible. This is because the following two properties apply.

Property 9: *An area used only for residential purposes is never situated between two mixed areas. (For a detailed proof, see Sasaki, 1990.)*

Property 10: *If there is a non-mixed residential area and all residents there commute inward, then no mixed area can be formed contiguously to the left of that residential area. (For the proof, see Sasaki. 1990.)*

Properties 9 and 10 restrict the possible incompletely mixed patterns to the two cases in Figure 12.10. The equilibrium conditions are deri-

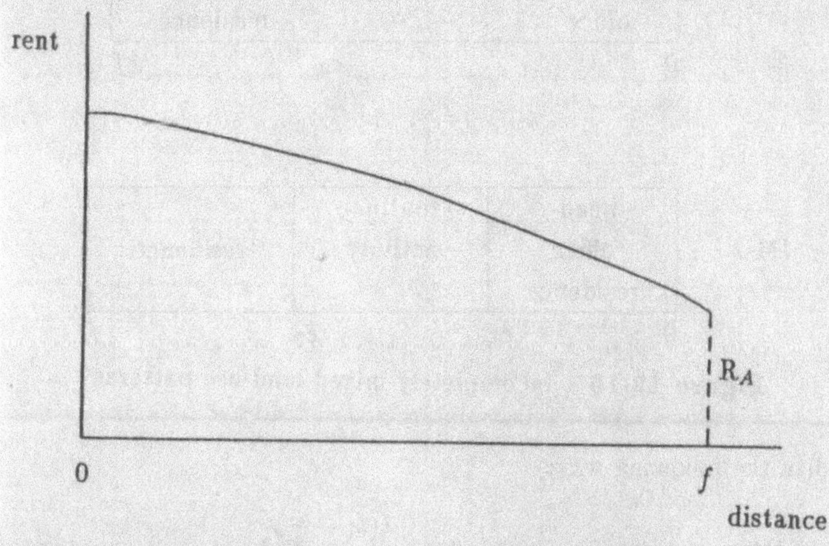

Figure 12.8 : Land rent structure of CM-1

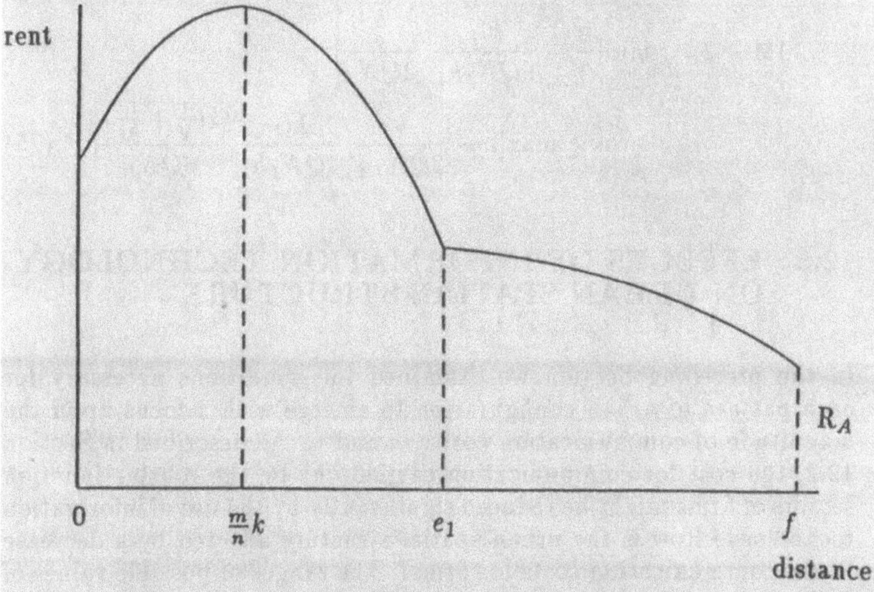

Figure 12.9 : Land rent structure of CM-2

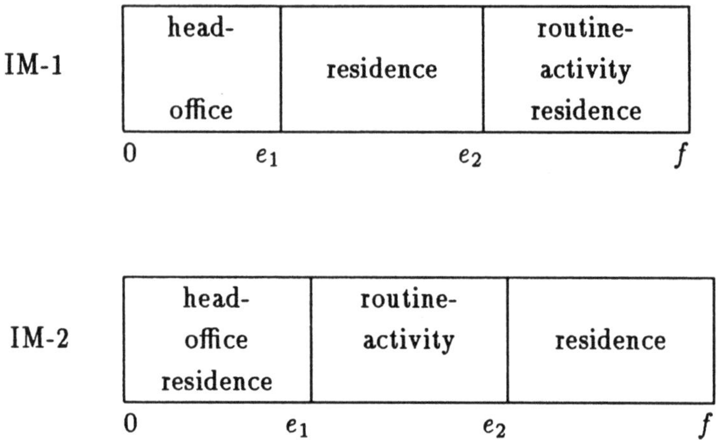

Figure 12.10 : Incompletely mixed land use patterns

ved in the following way:

$$\text{IM}-1 \quad \min\left[\frac{1}{h_0+l_0 q}\left\{nh_0 - \frac{k(l_0+\frac{h_0}{q})}{2\bar{Q}N_f}\right\}, \frac{k(l_1+\frac{h_1}{q})}{2\bar{Q}N_f\{2(h_0+l_0 q)+h_1+ql_1\}}\right] \geq m \tag{24}$$

$$\text{IM}-2 \quad \min\left[\frac{n}{3}+\frac{kl_1}{6\bar{Q}N_f h_1}, \frac{\frac{k}{q}}{2\bar{Q}N_f}\right] \geq m \geq \max\left[n-\frac{k}{2\bar{Q}N_f q}, \frac{kl_1}{4\bar{Q}N_f h_1}, \frac{k(\frac{1}{q}+\frac{l_1}{h_1})}{6\bar{Q}N_f}\right]. \tag{25}$$

12.5 EFFECTS OF INFORMATION TECHNOLOGY ON URBAN SPATIAL STRUCTURE

In the preceding Section we examined the conditions necessary for each pattern of urban configuration to emerge with a focus upon the magnitude of communication costs, m and n. As described in Section 12.2, the cost for communication carried out by the routine-function sectors of firms might be reduced significantly by the use of information technology. How is the urban spatial structure affected by a decrease in the communication costs for firms? The ranges of possible values of m for each configuration pattern, which were investigated in detail in Section 12.4, can suggest the answers to this question.

Before referring to the effects on urban spatial structure, we will specify the relation between the pervasion of information-communication technology and the values of m. If all the trips for communications from a firm's routine-function sector to its head-office and the trading floor are substituted by telecommunication methods, those communication costs cease to depend on the distance from the routine-function sector, although the cost by particular methods, e.g., telephone, might to some extent depend on the distance.

However, we do not imagine a situation where all the trips for communications made by the routine-function sector are substituted by information technology, but rather a situation where the ratio of trips substituted increases with the pervasion of information technology but some portions of communications are still carried on in face-to-face meetings. This assumption is not unrealistic: reporting to the head-office on business and exchanging views on the plans for management, production, sales, and inventories will be made in face-to-face meetings even after the information technology pervades firms. In this sense, the communication cost per unit distance m is regarded as decreasing with the pervasion of information technology. Before 'informationization', when most communications are carried on in face-to-face meetings and hence m is rather large, the EX-1 configuration emerges. In other words, in this situation a firm tends to be located in the central area of a city without separating the routine-function sector from its head-office. On the other hand, when 'informationization' develops so that most communications of the routine-function sector are carried out by information technology and thereby m is rather decreased, the CM-1 pattern is realized. An interesting observation is that firms are decentralized, but their routine-function offices are not physically separated from their head-offices in the CM-1 configuration which emerges in the highly 'informationized', society. Therefore, in this situation, the distance of trips made by firms' head-offices for face-to-face contacts on the trading floor will be rather longer than those in the 'Pre-informationization' situation (EX-1 pattern), given the density of firms and households.

We should note that CM-1 is realized when the cost for face-to-face communications carried out by the head-office, n is rather small and there is a completely mixed pattern. This pattern does not occur where n is to some extent large, even though m is decreased as the information technology pervades a society.[7]

It is when the sum of m and n is of medium size that the routine-activity office is located in the peripheral area of a city separated from

[7] As described in footnote 6, n reflects the time cost of executives, and hence a particular type of improvement of the transportation system, e.g., increased speed will decrease n to some extent.

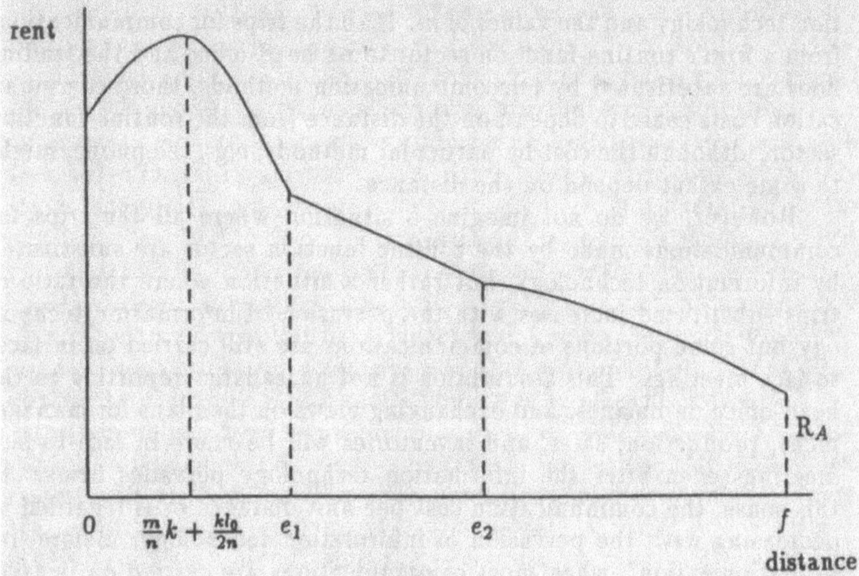

Figure 12.11 : Land rent structure IM-1

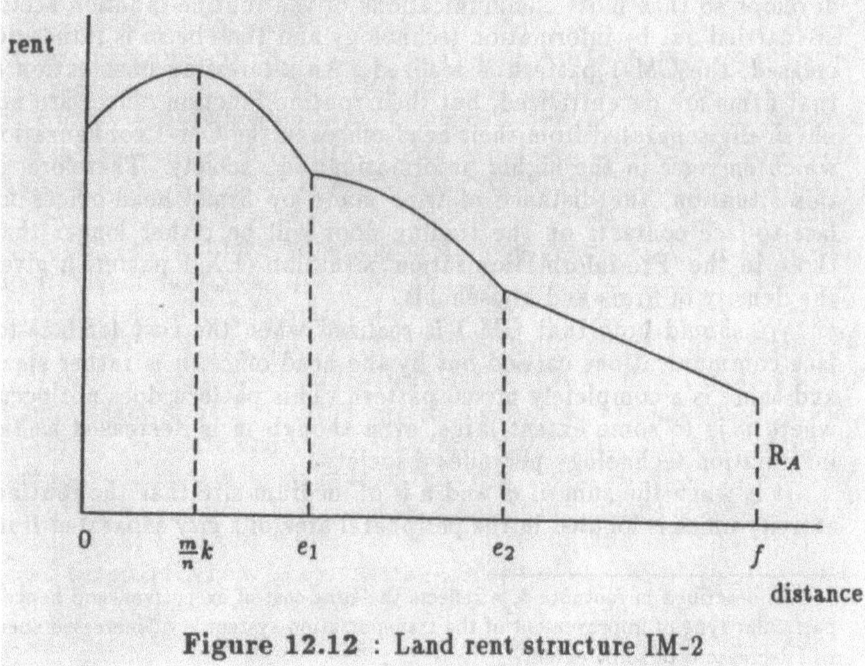

Figure 12.12 : Land rent structure IM-2

the head-office. If m is decreased by the introduction of information technology, but is larger than some lower limit, and n is still large, then one of EX-2, EX-3, EX-4, EX-5, or IM-2 will possibly emerge. If m is further decreased, then it is predicted that CM-2 or IM-1 will occur.

We note that the sizes of m and n have not been referred to in terms of *absolute* values, but rather *relative* values in relation to other parameters. Among the important parameters is the commuting cost per unit distance k, which depends on the urban transportation system. Even though m is decreased with 'informationization', firms are not decentralized but rather centralized if k is decreased at a rate higher than that of m as a result of an improved transportation system.

Acknowledgement: This research was supported by grants from the Telecommunications Advancement Foundation and the Japan Security Scholarship Foundation, which the author gratefully acknowledges.

REFERENCES

Evans, A.W., 1985, *Urban Economics: An Introduction,* Basil Blackwell, Oxford.

Ogawa, H. and M. Fujita, 1978, "Land Use Pattern in a Nonmonocentric City", Working Papers in Regional Science and Transportation No.8, University of Pennsylvania.

Ogawa, H. and M. Fujita, 1980, "Equilibrium land use patterns in a nonmonocentric city", *Journal of Regional Science,* 20:4:455-475.

Pressman, N.E.P., 1985, "Forces for Spatial Change", in J. Brotchie, P. Newton, P. Hall and P. Nijkamp (eds.) *The Future of Urban Form,* Croom Helm (Reprinted by Routledge 1989).

Sakamura, K., 1989, *Denno Mirai Ron,* Kadokawa-Shoten, Tokyo (in Japanese).

Salomon, I., 1986, "Telecommunications and travel relationships: A review", *Transportation Research A,* 20A:3:223-238.

Sasaki, K., 1990, "Information Technology and Urban Spatial Structure", Discussion Paper Series No. 29 Research Center for Applied Information Sciences, Tohoku University.

CHAPTER 13

Impacts of Developments in Telecommunication Systems on Travel Demand and the Location of Office Firms

Se-il Mun

13.1 INTRODUCTION

Office firms must communicate with other firms, mainly via face-to-face contacts. It has been understood that the need for such contacts is the driving force behind the agglomeration of office firms in the downtown area of cities; see, for example, O'Hara, (1977); Fujita and Ogawa, (1982); Tauchen and Witte, (1983, 1984).

Communications with other firms can also be conducted by telephone, or more advanced telecommunication means such as facsimile, electronic mail, videophones, etc. whose availability and convenience have grown rapidly in recent years. It is sometimes claimed that future developments in information technology may drastically affect the form and location patterns of office work. For example, it is predicted that most people will be working in their homes with computer terminals and all communications among firms will be made by telecommunications; hence, there will be no reason for the agglomeration of firms in the central business districts.

These views overlook an important factor in communication; that is, as Evans (1985) suggested, the quality of information. The quality of information in face-to-face contacts is different from that in various means of telecommunications. Means of telecommunications take advantage of improvements in the methods of storing, processing and transfering standardized information. On the other hand, face-to-face contacts are mainly conducted by elite executives and they deal with unstandardized, complex problems such as policy making, negotiations, etc., which are difficult to deal with via telecommunications. Although further developments in information technology might be able to improve the handling of higher quality information, it is also anticipated that the increasing complexity of the socio-economic system will necessitate even more face-to-face communication for high quality information. Hence, only *partial* substitution may occur between face-to-face contacts and telecommunications, and the importance of face-to-face contacts will not decrease. It is, however, still unclear whether travel demand will increase or decrease, and how the urban structure will be changed by the development of telecommunications technology. Salomon (1985) has discussed the relationship between travel and telecommunications. He suggests that increased interaction through telecommunications is likely to enhance the need for face-to-face interactions, that is, to generate new travel demand. Therefore, even though current travel patterns are reduced by substitution, the net effect on the number of trips may be close to zero.

These issues have seldom been studied using analytical models. Recently, Sasaki (1990a) and Ota and Fujita (1990) analyzed the urban spatial structure in which it becomes possible for firms to operate their activities in multiple locations and to conduct intra-firm communications via telecommunications. Further, Sasaki (1990b) has analyzed the case of housewives who work at home with computer terminals while their husbands commute to their offices in the CBD. The main focus in these studies is on spatial configurations of different land uses, so the impacts on the physical structure of urban areas and travel demand are not analyzed. Kobayashi and Okada (1989) consider the accessibility to communications by both travel and telecommunications as important factors in knowledge production. Although they analyzed the effects of changes in the telecommunication system on the location of firms in the regional system, the effects on travel demand were not analyzed. Their model cannot describe the formation of communication demand by all modes since they used accessibility measures.

The aim of this Chapter is to analyze the impacts of telecommunications developments on travel demand and the location of office firms. We formulate a model of office firms in which the choice between telecommunications and face-to-face contacts is described by incorporating the quality of information in communication activities.

13.2 THE MODEL

Suppose that a city consists of I discrete zones ($i = 1, 2, \cdots, I$). Each zone is assumed to be homogeneous and has identical size. The distance between a firm in zone i and any other firm in zone j is represented by the distance between the centres of each zone, t_{ij}. There exist TN office firms in the city, each of which has identical technology.

13.2.1 The Quality of Information and Communication Costs

There are numerous purposes for communications between firms, such as requesting or providing information and services, the exchange of ideas, negotiations, conferences, etc., and they range from the simple to the complicated, from the routine to the unstandardized. Let us assume that these various types or levels of communication can be characterized by a single variable, q, namely *the quality of information.* By using this variable, we can say that requesting or providing information is low level communication, while conferences are of a high level, because the latter take longer and deal with unstandardized issues, and so on.

Further, the frequency of communications between any pair of firms is not limited. It is natural to suppose that each firm conducts several rounds of communications, each of which may be at a different quality level. For example, to complete one contract, they may require some meetings for instruction, some for negotiation, and some for signing the contract, etc. Let us introduce a variable, $n(q)$, which represents the demand for communications at each quality level. $n(q)$ is defined as the density function, and it is assumed to be distributed over the range (q_L, q_U), where q_L and q_U are the lower and upper bounds of the level of the quality of communication respectively. Thus $n(q)$ must satisfy the following condition,

$$\int_{q_L}^{q_U} n(q)dq = 1. \tag{1}$$

The form of $n(q)$ depends on the characteristics of the activities in each type of firm, and is given exogenously in our analysis.

Two types of communication modes are considered, i.e. face-to-face contacts and telecommunications. Let us denote by $c_{ij}^h(q)$ the cost for communications at quality level q from a firm in i to a firm in j, by mode h, $h = 1,2$, where $h = 1$ and 2 correspond respectively to face-to-face contacts and telecommunications. These costs are specified as follows:

$$c_{ij}^1(q) = et_{ij}, \tag{2}$$

$$c_{ij}^2(q) = ft^2(q), \tag{3}$$

where e represents the monetary and time costs per unit distance of travel, t_{ij} is the distance between zones i and j. f represents the fees for the use of telecommunications facilities per unit of time, $t^2(q)$ is the time required to conduct communication at level q, which is an increasing function of q. Note that $c_{ij}^2(q)$ does not depend on the distance to the other firms, as intra-city telephone charges are adopted.

The choice of mode for communication is described as follows. Each firm chooses the mode h for communication at level q if

$$c_{ij}^h(q) = \min\{c_{ij}^1(q), c_{ij}^2(q)\}. \tag{4}$$

The average cost for communication TC_{ij} can be written

$$TC_{ij} = \int_{q_L}^{q_U} n(q) \min\{c_{ij}^1(q), c_{ij}^2(q)\} dq. \tag{5}$$

It is assumed that $c_{ij}^1(q_L) > c_{ij}^2(q_L)$ and $c_{ij}^1(q_U) < c_{ij}^2(q_U)$. This assumption guarantees that the higher (lower) level communication is necessarily conducted by face-to-face contacts (by telecommunications), as discussed in Section 13.1. Then, TC_{ij} can be rewritten as

$$TC_{ij} = \int_{q_L}^{q_{ij}^*} n(q) c_{ij}^2(q) dq + \int_{q_{ij}^*}^{q_U} n(q) c_{ij}^1(q) dq, \tag{6}$$

where q_{ij}^* is determined so that the following relation holds

$$c_{ij}^1(q_{ij}^*) = c_{ij}^2(q_{ij}^*), \tag{7}$$

that is, q_{ij}^* is determined so that the costs for communication by each mode are indifferent at q_{ij}^*.[1]

By using the specifications in (2) and (3), (7) can be rewritten as

$$et_{ij} = ft^2(q_{ij}^*). \tag{8}$$

Communication at a lower level than q_{ij}^* is conducted by telecommunication systems, while that higher than q_{ij}^* is conducted by face-to-face contacts. Thus, the share of face-to-face contacts for communication between i and j is calculated by

$$Q_{ij} = \int_{q_{ij}^*}^{q_U} n(q) dq. \tag{9}$$

[1]If $c_{ij}^1(qL) < c_{ij}^2(qL)$, $q_{ij}^* = qL$. That is, all communications are conducted by face-to-face contacts.

13.2.2 Behaviour of a Firm

Let us first consider the outcome of communication activity. By communicating with other firms, each firm gets something which contributes to its production, such as information, clients, etc. We call this a 'benefit from communication' and denote it by V_i. V_i is represented as follows:

$$V_i = \sum_{j=1}^{I} \sum_{n=1}^{N_j} v(x_{ij}^n), \tag{10}$$

where N_j is the number of firms in j, x_{ij}^n is the amount of communication from a firm in i to a firm n in j, and $v(\cdot)$ is an increasing and concave function. (10) implies that the firm can gain greater benefits if it communicates with many different firms. The concavity of $v(\cdot)$ implies that the marginal gain from contacts per firm is decreasing since the benefits (such as information) obtainable from any individual source are limited.

If we assume that it is inconsequential which firms in j a firm in i communicates with, then

$$x_{ij}^1 = x_{ij}^2 = \cdots = X_{ij}/N_j,$$

where X_{ij} is the total amount of communication from a firm in i to firms in j, that is

$$X_{ij} = \sum_{n=1}^{N_j} x_{ij}^n. \tag{11}$$

Thus (10) is rewritten as

$$V_i = \sum_j N_j v(\frac{X_{ij}}{N_j}). \tag{12}$$

It is assumed that the inputs for production of a firm are V_i and the labour for routine work, Z_i. The production function is $F(V_i, Z_i)$. Also, each firm occupies a certain area of office space, G, which is assumed to be a given constant. The profit of a firm is defined by

$$\pi_i = pF(V_i, Z_i) - wZ_i - r_iG - \sum_j X_{ij}TC_{ij}, \tag{13}$$

where p, w, and r_i are the price of the products, the wage rate, and the rent per unit floor space, respectively. Each firm chooses the level of communication X_{ij}, the mode for communication, the level of routine work Z_i, and its location i, so as to maximize its profit. Recall that the choice of communication mode has been explained in Section 13.2.1.

The first-order conditions are as follows:

$$p\frac{\partial F}{\partial V_i}\frac{\partial V_i}{\partial X_{ij}} - TC_{ij} = 0 \tag{14}$$

$$p\frac{\partial F}{\partial Z_i} - w = 0. \tag{15}$$

To solve explicitly the conditions (14) and (15), we specify the form of the functions as follows

$$v(X_{ij}/N_j) = (X_{ij}/N_j)^{\sigma}, \qquad 0 < \sigma < 1 \tag{16}$$

$$F(V_i, Z_i) = V_i^a Z_i^b, \qquad a + b = 1, \quad a, b > 0. \tag{17}$$

Incorporating these specifications into (14) and (15), and by manipulation, we obtain the following solutions

$$X_{ij}^* = \{(\sigma a)p^{\frac{1}{a}}(\frac{w}{b})^{-\frac{b}{a}}\}^{\frac{1}{1-\sigma}} N_j TC_{ij}^{\frac{1}{\sigma-1}} \tag{18}$$

$$Z_i^* = \{(\sigma a)^{\sigma} p^{\frac{1}{a}}(\frac{w}{b})^{\frac{\sigma a-1}{a}}\}^{\frac{1}{\sigma-1}}(\sum_j N_j TC_{ij}^{\frac{\sigma}{\sigma-1}}). \tag{19}$$

Let us define the total number of communications conducted by each firm in i, O_i, which is obtained by

$$O_i = \sum_j X_{ij}. \tag{20}$$

Then, (18) is rewritten as follows

$$X_{ij} = O_i \frac{N_j TC_{ij}^{\frac{1}{\sigma-1}}}{\sum_k N_k TC_{ik}^{\frac{1}{\sigma-1}}} = O_i P_{ij}. \tag{21}$$

Note that $1/(\sigma - 1)$ takes a negative value since $0 < \sigma < 1$. (21) is a gravity model, which is derived from the conditions for profit maximization of a firm.[2]

[2] If we adopt the specification of $v(Xij/Nj) = (Xij/Nj)ln(Xij/Nj)$, the negative exponential type gravity model,

$$X_{ij} = \frac{N_j \exp(-TC_{ij})}{\sum_k N_k \exp(-TC_{ik})}$$

is derived. However, this contradicts the assumption that $v(\cdot)$ is strictly increasing.

13.2.3 Spatial Equilibrium of Office Locations

The conditions for spatial equilibrium are essentially the same as in Mun and Yoshikawa (1993). In our model, the total number of firms is fixed by TN. From the assumption of identical firms, all firms achieve the same profit level π^*, regardless of their location in a city. Then, if we denote an equilibrium distribution of firms by $(N_1^*, N_2^*, \ldots N_I^*)$, the equilibrium conditions are written as

$$\pi(r_i^* \mid \{X_{ij}^*, q_{ij}^*, j = 1, I\}, Z_i^*) = \pi^*, \quad \text{if } N_i^* > 0 \tag{(22)a}$$

$$N_i^* = 0, \quad \text{if } \pi(r_i^* \mid \{X_{ij}^*, q_{ij}^*, j = 1, I\}, Z_i^*) \leq \pi^* \tag{(22)b}$$

and

$$\sum_i N_i^* = TN \tag{23}$$

where

$$\begin{aligned} &\pi(r_i^* \mid \{X_{ij}^*, q_{ij}^*, j = 1, I\}, Z_i^*) \\ &= p\{\sum_j N_j^{*1-\sigma} X_{ij}^{*\sigma}\}^a Z_i^{*b} - wZ_i^* - r_i^* G - \sum_j X_{ij}^* TC_{ij} \end{aligned} \tag{24}$$

$$r_i^* = \beta c (N_i^* G / LA_i)^{\beta - 1}. \tag{25}$$

LA_i is the land area used for office buildings in zone i, β and c are constants. The equilibrium rent r_i^* is derived by the same method as that in Mun and Yoshikawa (1993).

13.3 THE SHORT-RUN EFFECTS OF DEVELOPMENTS IN TELECOMMUNICATION SYSTEMS

In this Chapter, developments in telecommunication systems are represented by a decrease in f, the fees for the use of telecommunication facilities per unit of time. A decrease in $t^2(q)$, namely savings in the time required to complete communication at a certain quality level q, may also occur due to the development of telecommunication systems with higher performance. But this has qualitatively the same effects as a decrease in f. First we analyze the impacts on travel demand in the situation in which the location of firms is not changed. These can be regarded as short-run effects. And later, we analyze the long-run effects in situations where the location may be changed. Furthermore, we also compare the long-run effects on travel demand with short-run effects.

Suppose that spatial equilibrium is currently achieved in the city. Travel demand for face-to-face contacts from i to j is computed by $N_i^* X_{ij}^1$, where X_{ij}^1 is the amount of communication by face-to-face contacts from a firm in i to firms in j, that is, from (9)

$$X_{ij}^1 = X_{ij} Q_{ij}. \tag{26}$$

Since N_i^* is unchanged in the short run and all firms are identical, the effects of a decrease in f are analyzed by differentiating (26) with respect to f

$$\frac{\partial X_{ij}^1}{\partial f} = \frac{\partial X_{ij}}{\partial f} Q_{ij} + X_{ij} \frac{\partial Q_{ij}}{\partial f}, \tag{27}$$

where

$$\frac{\partial X_{ij}}{\partial f} = \{(\sigma a) p^{\frac{1}{a}} (\frac{w}{b})^{-\frac{b}{a}}\}^{\frac{1}{1-\sigma}} \frac{1}{\sigma - 1} N_j TC_{ij}^{\frac{2-\sigma}{\sigma-1}} \frac{\partial TC_{ij}}{\partial f} < 0 \tag{28}$$

since

$$\frac{\partial TC_{ij}}{\partial f} = \int_{q_L}^{q_{ij}^*} n(q) t^2(q) dq > 0 \tag{29}$$

and

$$\frac{\partial Q_{ij}}{\partial f} = -n(q_{ij}^*) \frac{\partial q_{ij}^*}{\partial f} > 0 \tag{30}$$

since

$$\frac{\partial q_{ij}^*}{\partial f} = -\frac{t^2(q_{ij}^*)}{f t^{2\prime}(q_{ij}^*)} < 0 \quad \text{from} \quad (8). \tag{31}$$

The first term in the R.H.S. of (27) is negative, and the second term is positive. That is, the total demand for communication X_{ij} is increased by a decrease in f while the share of face-to-face contacts Q_{ij} is necessarily decreased. This implies that the impact of telecommunication costs on travel demand is ambiguous. Furthermore, from (31), the level of communication at which telecommunication systems are chosen becomes higher.

To see the effects of the change in f more comprehensively, numerical analyses are conducted. To do so, the function $n(q)$, the frequency of communication at each quality level, q, is specified as follows,

$$n(q) = A q^{-B} \quad A, B > 0 \tag{32}$$

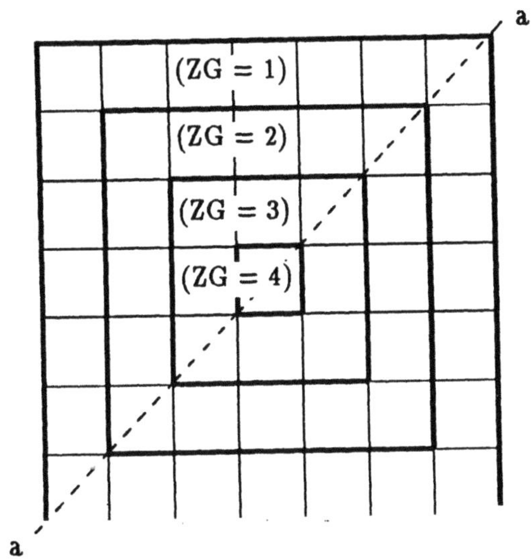

Figure 13.1 : Numerical analysis of the city

Table 13.1 : Parameter values for the numerical analysis

p=1.7	a=0.45	w=24
e=0.3	G=114.5	β=1.5
c=0.6	LA_i=100000	(i=1,..,49)
B=1.5	τ=1.0	
$q_L = 0.1$	$q_U = 10.0$	

where B is a constant, A is determined so as to satisfy the condition (1). This specification implies that the frequency of communication is decreased as its level of quality becomes higher, which is consistent with the observation by Goddard (1973).

In the numerical analysis, the city is divided to 49 zones as shown in Figure 13.1. The zones are classified into four rings by distance from the city centre; see $ZG = 1$ to $ZG = 4$ in Figure 13.1. The parameter values used in numerical analysis are listed in Table 13.1. The value

Table 13.2 : The values of DX for various levels of telecommunication costs with origin-destination pairs

	Case A : σ=0.30			Case B : σ=0.34			Case C : σ=0.40		
	Level of Telecommunication Costs			Level of Telecommunication Costs			Level of Telecommunication Costs		
	high	medi.	low	high	medi.	low	high	medi.	low
Centre-Periphery	–	0	++	0	–	++	–	–	++
Periphery-Centre	0	0	+++	–	–	+++	–	– –	+++
Centre-Intermediate	+	0	–	+	0	– –	+	0	– –
Periphery-Intermediate	+	+	0	+	0	– –	+	–	– – –

Note: + or – indicates the sign of DX. The relative magnitude of DX is represented by the number of signs.
0 means that the magnitude of DX is very small, regardless of its sign.

of telecommunication cost f is confined to $0.3 < f < 1.4$, so as to guarantee $0 < Q_{ij} < 1$, for all i, j, which is assumed in Section 13.2.

Three cases are examined in the numerical analysis, i.e. Case A : $\sigma = 0.3$, Case B : $\sigma = 0.34$ and Case C : $\sigma = 0.4$.

In Table 13.2, the values of the L.H.S. in (27) (DX hereafter) for some representative origin-destination pairs are shown for various values of f. In the Table, 'Centre' and 'Periphery' mean the set of zones classified to $ZG = 4$ and $ZG = 1$, respectively, and 'Intermediate' means those in $ZG = 2$ and 3. The levels of telecommunication costs represent the stages of technological development: for example, high costs for telecommunications imply an early stage in technological development, since it is expected that telecommunication costs are reduced by technological development. Table 13.2 shows that the values of DX are either positive or negative. The sign of DX depends on the O-D pair, the values of σ and the stages of technological development. For O-D pairs of a short (long) distance, the value of DX is negative (positive) when f is large, and positive (negative) when f is small. This implies that the travel demand for longer distances is decreased by a decrease in telecommunication costs at the beginning of telecommunication development, while that for shorter distances is increased. Next, as σ becomes larger, the range of f where DX is negative becomes larger. In other words, travel demand tends to be increased as the value of parameter σ increases. The parameter σ can be interpreted to represent how actively the firm seeks diversity in its communication activity. As σ becomes lower, each firm seeks more diversity in communication. As a result, the firm tends to have contact with firms locating in more distant zones.

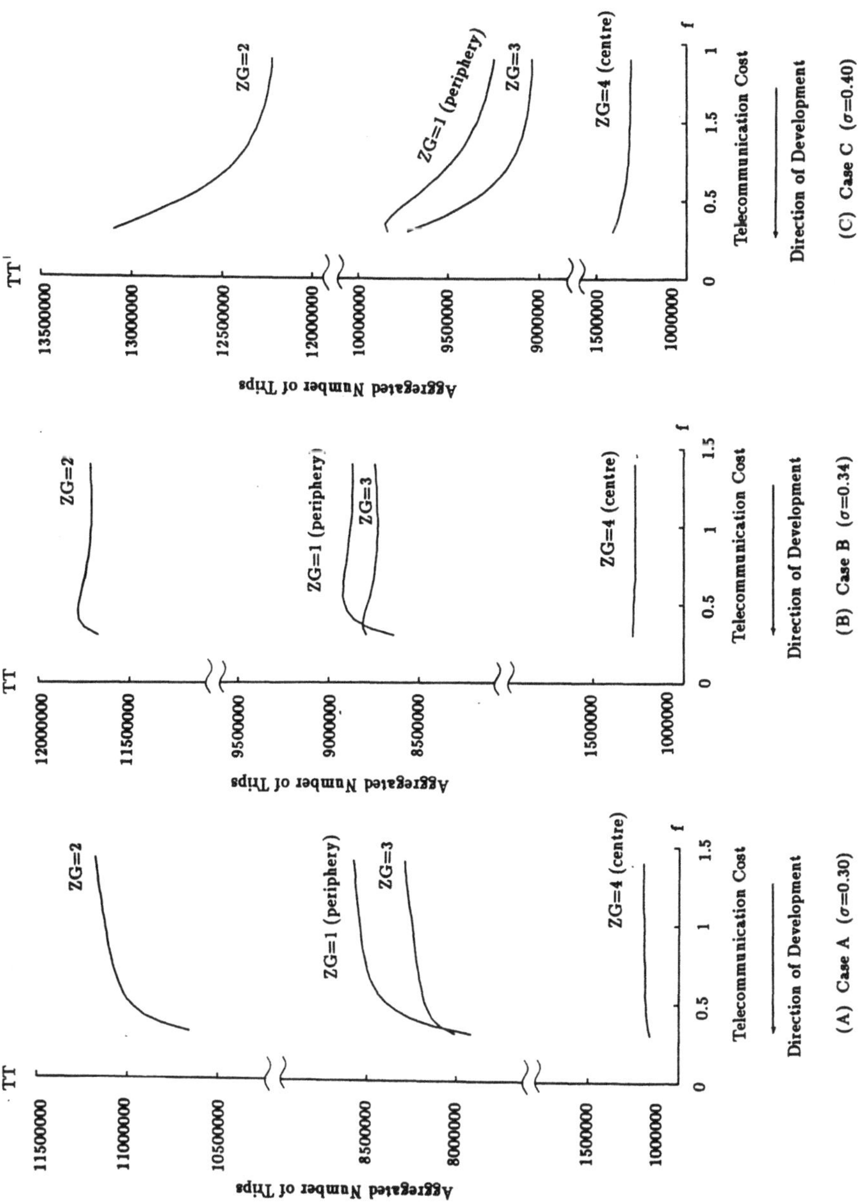

Figure 13.2 : Telecommunication costs and the number of trips aggreagted by rings

The effects on the aggregated number of trips are shown in Figures 13.2 (A)-(C). The vertical axis in each figure is the total number of trips originating from each ring, which is calculated by

$$TT_{ZG} = \sum_{i \in \Omega_{ZG}} N_i \sum_j X_{ij} Q_{ij} = \sum_{i \in \Omega_{ZG}} N_i O_i^1 \tag{33}$$

where Ω_{ZG} means the set of zones in the ring ZG, and O_i^1 is the number of trips from a firm in i. For Case A (Figure 13.2 (A)), the number of trips from each ring is necessarily decreased by a decrease in f, while the number is increased for case C (Figure 13.2 (C)). The results for Case B (Figure 13.2 (B)) are different; at the outer ring, the number of trips is increased by a decrease in f when f is high, and the number starts to decrease if f is lower than 0.6. On the other hand, in the inner rings, the number of trips is decreased when f is high, increased for intermediate values of f, and sharply decreased when f is low. As a whole, the number of trips is apt to be increased in inner locations, at the later stages of telecommunication development (when f is low), and as firms seek less diversity in communication activities (when σ is high).

13.4 THE LONG-RUN EFFECTS ON LOCATION AND TRAVEL DEMAND

In the long-run, firms will change their location in response to changes in their environment, and a new equilibrium distribution of firms will be achieved. In this Section, we analyze the effects of change in f on the spatial distribution of office firms and travel demand.

As in the analysis of short-run effects, the following three cases are examined; i.e., Case A : $\sigma = 0.3$, Case B : $\sigma = 0.34$, Case C : $\sigma = 0.4$.

Let us begin with the impacts on the spatial distribution of firms' locations. Figure 13.3 shows the spatial distribution of office firms for Case A: the number of firms in the zones on line $a - a$ in Figure 13.1 are plotted. It is seen that the number of firms in the inner zones is decreased by a decrease in f, while the number of firms in the outer zones is increased. That is, spatial distribution is monotonously decentralized by a decrease in f. The results are the same for Cases B and C.

Let us move to the impacts on travel demand. Recall (33), the definition of the aggregate number of trips. In the long run, unlike in the short run, both N_i and O_i^1 are changed. Hence, the impacts on the aggregated number of trips are divided into two parts; i.e., the effects of location change and travel demand from individual firms. We will see the effects on the aggregated number of trips after the effects on

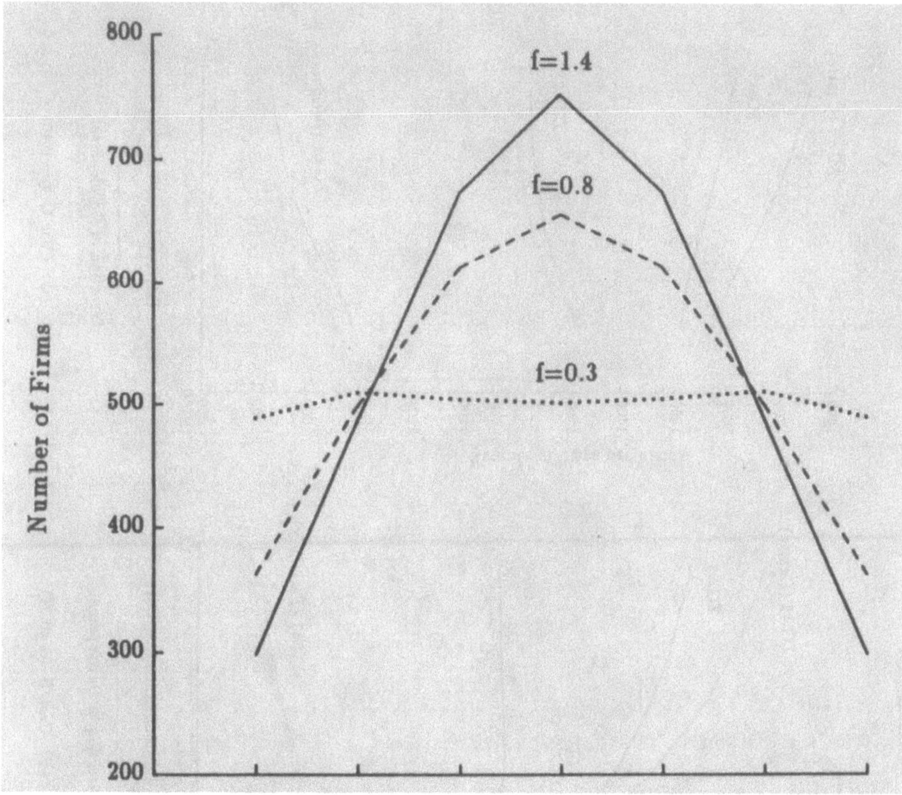

Figure 13.3 : Telecommuniation costs and the spatial distribution of firms

O_i^1 are investigated. Figures 13.4 (A)-(C) show the relation between telecommunication costs and the number of trips per firm in each ring. For Case A, the number of trips is decreased by a decrease in f except for the periphery ring. As σ becomes larger, the number of trips in the outer area increases, but the number of trips in the central zone is always decreased.

The effects of the reduction of telecommunication costs on the aggregated number of trips in each ring are summarized in Table 13.3. The short run effects are also shown for comparison. The long-run effects are qualitatively the same for all cases; the number of trips is increased in the periphery ring and decreased in the other rings. The effects of the decentralization of firms are dominant.

Comparing the long-run effects with the short-run effects, the impacts are completely the opposite in some cases; for example, at $ZG = 1$ in Cases A and B, and at $ZG = 3$ in Case C, etc.

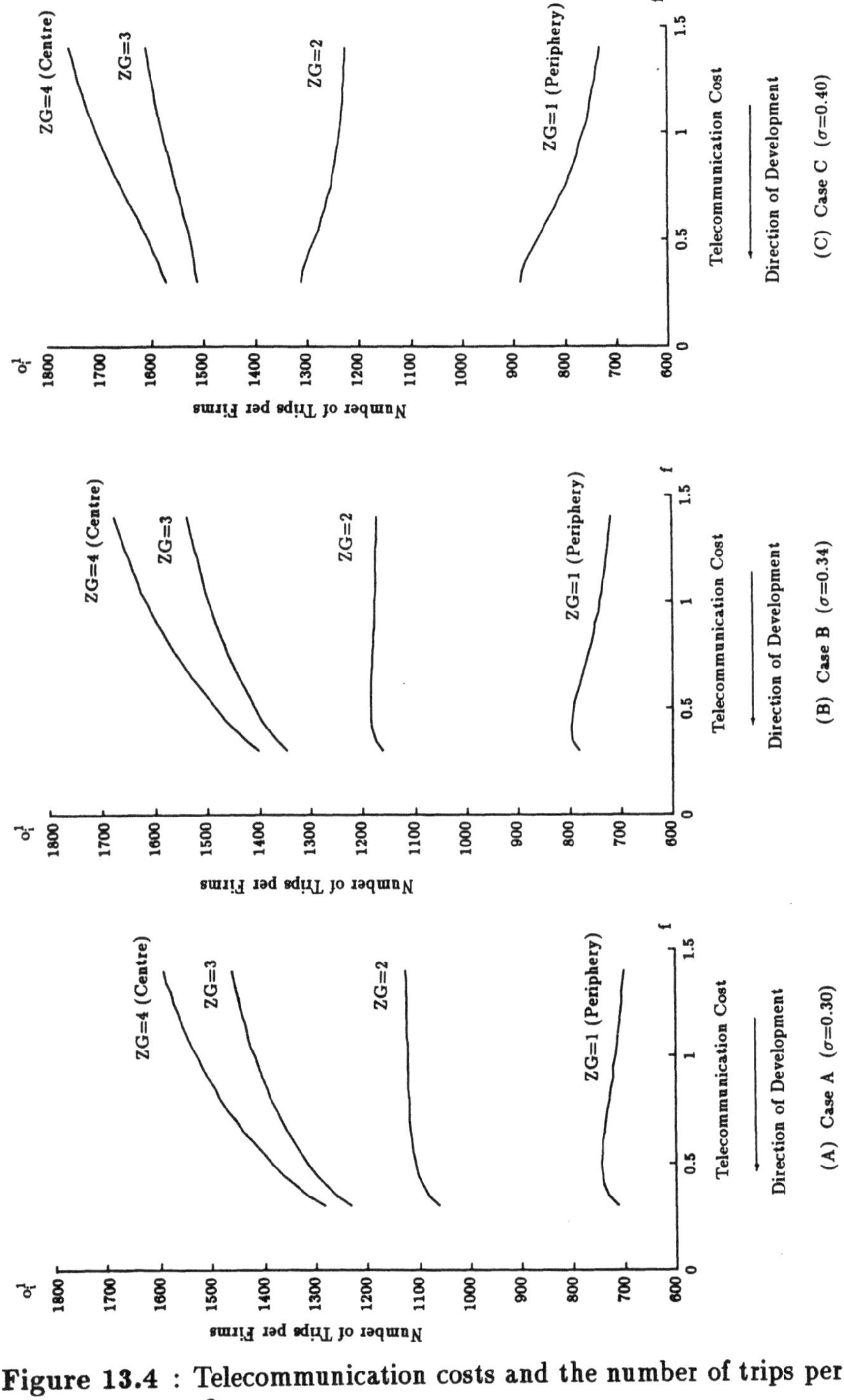

Figure 13.4 : Telecommunication costs and the number of trips per firm

Table 13.3 : The effects of decreasing telecommunication costs on travel demand in each ring (comparison between short run and long run effects)

	Case A : σ=0.30			Case B : σ=0.34			Case C : σ=0.40		
	Level of Telecommunication costs			Level of Telecommunication costs			Level of Telecommunication costs		
	high	medium	low	high	medium	low	high	medium	low
ZG=4 (Centre)	▼ ▼	▼ ▼	▼ ▼	▼ –	▼ –	▼ –	▼ –	▼ –	▼ △
ZG=3	▼▼▼ ▼	▼▼▼ ▼	▼▼▼ ▼▼	▼▼▼ –	▼▼▼ –	▼▼▼ △	▼▼▼ –	▼▼▼ △△	▼▼▼ △△△
ZG=2	▼▼ ▼	▼▼ ▼	▼▼ ▼▼▼	▼▼ –	▼▼ △	▼▼ ▼	▼▼ △	▼▼ △△	▼▼ △△△
ZG=1 (Periphery)	△△ ▼	△△ ▼	△△ ▼▼▼	△△ –	△△ △	△△ ▼▼▼	△△ △	△△ △△	△△ –

Note : The long-run and short-run effects are shown in the upper and lower line of each cell respectively. △ and ▼ represent the increase and decrease of the trips due to decreasing telecommunication costs, respectively. The number of these symbols represents the magnitude of the changes in the number of trips. – means that the magnitude of the effect is very small.

13.5 THE IMPACT ON CITY SIZE

Thus far, the total number of office firms in the city is fixed and given exogenously. Under the assumption that the office sector is the only production activity in the city, the city size is equivalent to the total number of firms, which is TN on the R.H.S. of (23). Figure 13.5 illustrates the loci of the equilibrium profit which is determined by solving (22) and (23) for various values of TN. It is seen that, when the city size is small, the equilibrium profit increases as the city size expands. This implies that the agglomeration effects exceed the costs associated with the expansion of the city, such as increased rents and wages. The equilibrium profit reaches a peak at a certain level of city size. If the city expands further, the profit level begins to decrease.

Suppose that the city is a part of a large regional system, within which firms and households can freely relocate without costs. Firms enter the city as long as a positive profit is obtained there, while they withdraw if the profit falls below zero. In this setting, the equilibrium city size is determined by point o in Figure 13.5; here, no firm has an incentive to change its location since it cannot increase its profit by relocating elsewhere. Point Q in Figure 13.5 is also a candidate equilibrium point, but it is not stable. Hence we will not consider such points hereafter.

In Figure 13.5, two loci are drawn for different levels of telecommunication costs. If telecommunication cost, f, is lower, the locus of the

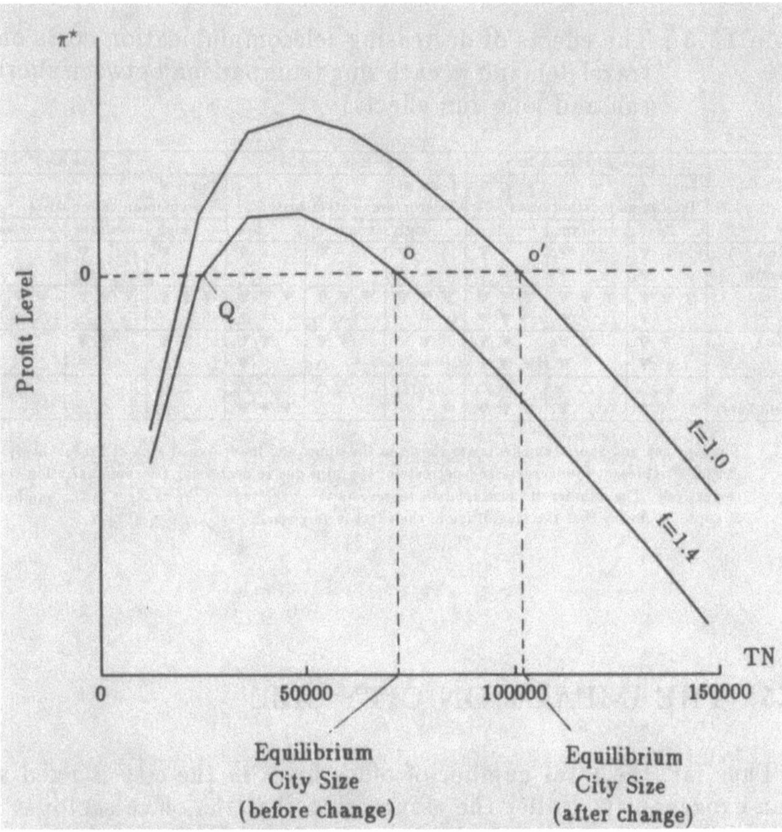

Figure 13.5 : Changes in city size due to decreasing telecommunication costs

equilibrium profit shifts upward. As a consequence, the equilibrium city size is expanded from point o to o'.

In Figure 13.6, the equilibrium city sizes are plotted for various levels of telecommunication costs. It is seen that the city size is monotonously decreasing as telecommunication costs increase: in other words, the city is expanding along with developments in telecommunication technology. Furthermore, the rate of change is larger when f is small; the more the telecommunication technology is advanced, the higher the speed of expansion.

The effects on travel demand and the spatial distribution of firms when the city size is variable are also investigated. The results are as follows: as f is reduced, the number of firms is increased in all zones, and the increment is larger in the outer zones. That is, the spatial distribution of the firms is relatively decentralized. Similarly, travel

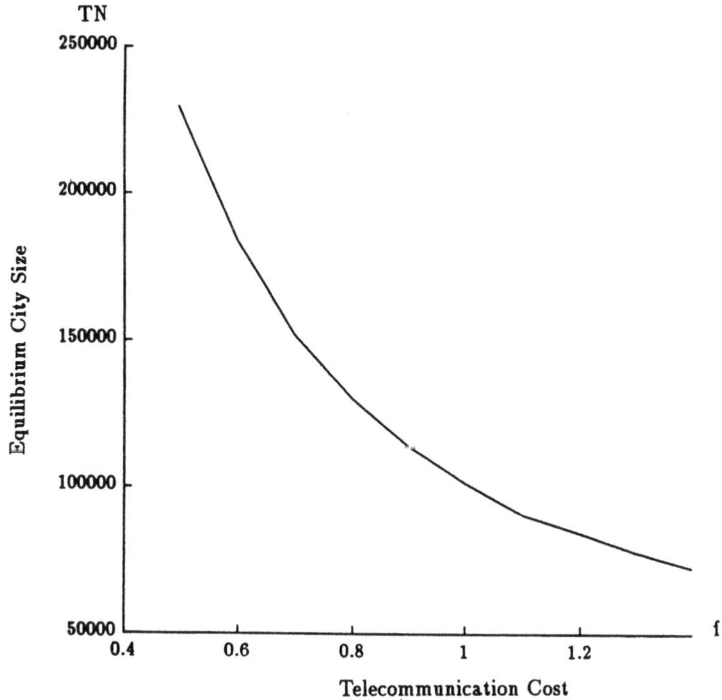

Figure 13.6 : Telecommunication costs and equilibrium city sizes

demand is increased in all zones, and the increment is larger in the outer zones.

13.6 SHIFT TOWARDS COMMUNICATION-INTENSIVE PRODUCTION

It is obvious that the role of communication in economic activities is becoming more and more important. The office sector we study is a typical activity to which this trend applies. This trend itself is not affected by the development of telecommunication technology, although it enhances the trend and affects the magnitude of change. We therefore focus on the shift in the production methods of office firms; we suppose that the value of parameter a in production function (17) rises as the role of communication becomes more important. Under the assumption of Cobb-Douglas technology, parameter a represents the contribution of increasing communication to the output of production. In this Section, we will consider only the long-run effects, since a parameter change takes a very long time.

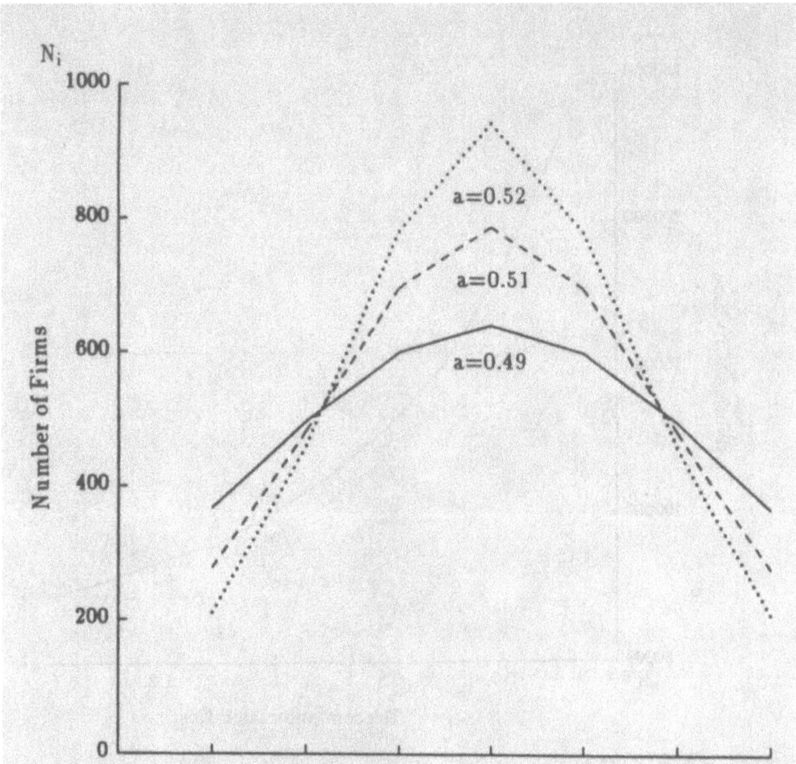

Figure 13.7 : The value of parameter a and the spatial distribution of firms

Figure 13.7 shows the spatial distribution of firms for various values of a. In the figure, as a becomes higher, the number of firms in the inner zones becomes larger and smaller in the outer zones. That is, the spatial distribution becomes concentrated by a shift toward communication intensive production. This result is in contrast to those concerning the effects of telecommunication cost reduction. If decreases in telecommunication costs and parameter shifts occur simultaneously, whether the firms are decentralized or concentrated depends on the relative strength of the two effects.

Figure 13.8 shows the relation between the number of trips per firm in each ring and the value of a. It is seen that travel demand increases as the value of a rises, and the increment is larger in inner locations. Incorporating these effects on the spatial distribution of firms, it is expected that the increase in aggregated trips in inner locations will be considerably large.

Figure 13.9 shows the loci of the equilibrium profit with changing city size for three different values of a. Since the locus of profit shifts upward as the value of a rises, the city size expands such that $o \rightarrow o'$,

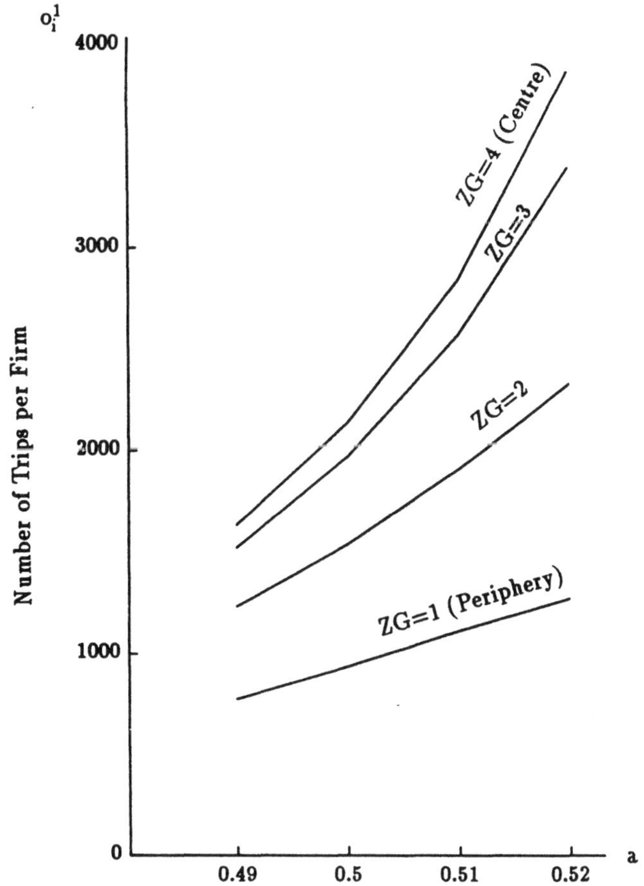

Figure 13.8 : The value of parameter a and the number of trips per firm

$o' \rightarrow o''$. The effect on the city size is similar to the effects of decreases in telecommunication costs. Therefore, if telecommunication cost reduction and parameter shifts occur simultaneously, the expansion of the city is accerelated.

13.7 CONCLUSION

In this Chapter, the effects of decreases in telecommunication costs on travel demand and the location of office firms are analyzed. Furthermore, the shift in the means of production toward communication-intensive technology is also analyzed.

Travel demand may be either increased or decreased by a decrease in telecommunication costs, depending on the parameter values and

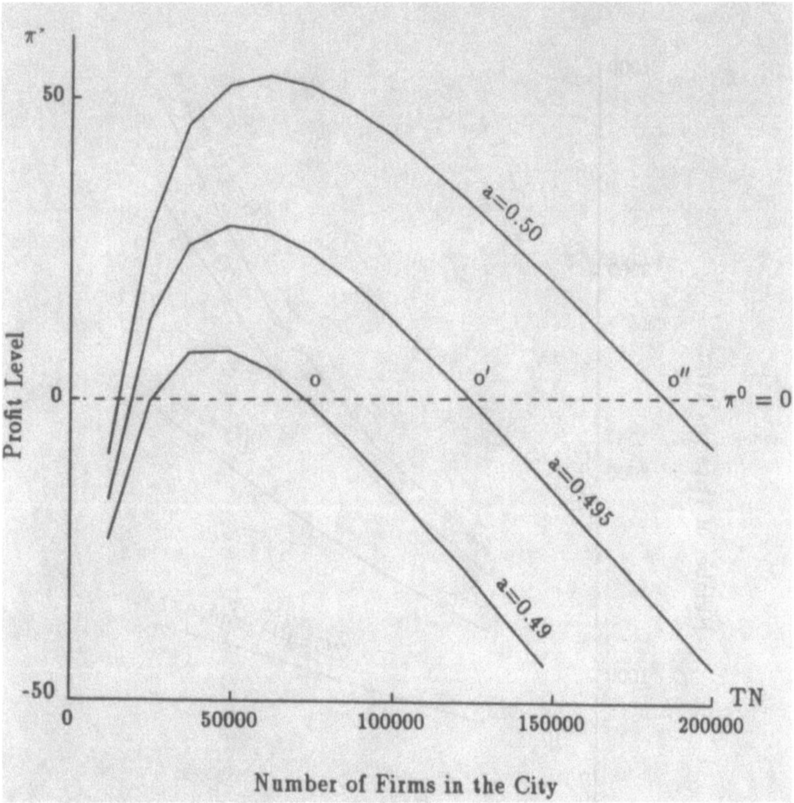

Figure 13.9 : Changes in city size due to changes in parameter a

stages of development in the telecommunication technology and the locations of the firms. Furthermore, in some cases, the short-run effects may be completely different to the long-run effects. Therefore, we must be very careful in planning investments in transportation infrastructure.

The spatial distribution of office firms within a city is decentralized by decreasing telecommunication costs, while the city size is expanded. On the other hand, if the means of production in the office firms becomes communication-intensive, the location of the firms becomes more concentrated and the city expands.

It is natural that decreases in telecommunication costs and parameter shifts occur simultaneously. In this case, the city size is necessarily expanded, but the impact on the intra-city spatial distribution is ambiguous. Empirical analysis is required to determine what occurs in reality.

REFERENCES

Evans, A.W., 1985, *Urban Economics; An Introduction*, Basil Blackwell, Oxford.

Fujita, M. and H. Ogawa, 1982, "Multiple equilibria and structural transition of non-monocentric urban configurations", *Regional Science and Urban Economics,* 12:161-196.

Goddard, J.B., 1973, "Office Linkages and Location, A study of communications and spatial patterns in Central London", *Progress in Planning*, Vol.1, Part 2, Pergamon.

Kobayashi, K. and N. Okada, 1989, "Technological Substitution between Telecommunications and Transportation in Production: A Theoretical Perspective", *Selected Proceedings of the 5th World Conference on Transport Research*, 2:241-256.

Mun, S. and K. Yoshikawa, 1993, "Communication among firms, traffic congestion and office agglomeration", forthcoming to *Annals of Regional Science*, Vol.26.

O'Hara, D.J., 1977, "Location of firms within a square central business district", *Journal of Political Economy,* 85:1189-1207.

Ota, M. and M. Fujita, 1990, "Communication Technologies and Spatial Organization of Multi-Unit Firms in Metropolitan Areas", unpublished paper, Department of Regional Science, University of Pennsylvania.

Salomon, I., 1985, "Telecommunications and travel, substitution or modified mobility?", *Journal of Transport Economics and Policy,* XIX:219-235.

Sasaki, K., 1990a, "Information Technology and Urban Spatial Structure", Research Center for Applied Information Sciences, Tohoku University.

Sasaki, K., 1990b, "The Effects of Information Technology on Urban Spatial Structure Where Home-Based Work Is Possible", Research Center for Applied Information Sciences, Tohoku University.

Tauchen, H. and A.D. Witte, 1983, "An equilibrium model of office location and contact patterns", *Environment and Planning A,* 15:1311-1326.

Tauchen, H. and A.D. Witte, 1984, "Socially optimal and equilibrium distribution of office activity; Models with exogenous and endogenous contacts", *Journal of Urban Economics,* 15:66-86.

CHAPTER 14

Spatial Equilibria of Knowledge Production with 'Meeting-Facilities'

Kiyoshi Kobayashi, Seishin Sunao, and Kazuhiro Yoshikawa

14.1 INTRODUCTION

One of the essential roles of the Agora in ancient Athens was to serve as a meeting place. The Agora consisted of an open square surrounded by a collection of buildings, which formed a focal point for the citizens' social interactions. The Agora was the civic centre for all. The activities surrounding the Agora, e.g., taverns, amusement places, etc. used to have social properties beyond their primary objectives. One of the central properties of the Agora was that of a collective good shared by all. The Agora, however, was not merely a public good. It was also a significant device to catalyze social interactions which facilitate the diffusion of new knowledge and information among citizens.

The cities of today, of course, possess much more affluent bundles of collective goods than those which ancient cities could have. The collective goods of modern cities form their so-called urban infrastructure. One of the clear characteristics of modern cities is that they have become the most prodigious memory-houses of knowledge that humankind has yet created. As an economy develops and as a society becomes more complex, efficient organizations of production, traders and governments seem to require an increasing degree of specialization and decentralization of knowledge. Eventually, nobody has control over access to the whole body of knowledge. The significance of collective assets covering the transmission, assimilation and accumulation of knowledge has increased to an extent never experienced by man before.

The tendencies toward decentralization and specialization in knowledge production inevitably bring about increasing needs for more efficient ways to transmit and exchange the knowledge which man has newly produced. Face-to-face contact between two individuals seems to be one of the most basic ways for bilateral exchanges of ideas, but it is not always efficient in the process of knowledge transmissions. 'Meetings', e.g., conferences, conventions, etc, are expected to reduce the frequency of human contacts by bringing together many persons in the same place, enabling them to have intense multi-person interactions during a limited time period. Thus, meetings are significant means for facilitating the diffusion of new ideas and knowledge. However, the utility of a meeting can grow to such a level that the number of participants generates congestion and it loses its economy of scale. Thus, meetings can be regarded as endogenous local public goods. We term such places accommodating any kind of meeting a 'meeting-facility'. Typical solutions to the infrastructure requirements for meetings are conference halls, hotels, and even cafes in modern cities. It is no exaggeration to say that the urban infrastructure which is designed to catalyze knowledge exchange among different urban groups forms an important ingredient in the very mechanism of change - growth processes - in modern cities.

Our emphasis will be on the interdependencies between stocks of knowledge and their distribution over space in the form of knowledge production firms which are located within different nodes of a discrete network. We shall also focus upon the essential role which meeting-facilities play in the process of knowledge exchange over a spatial network. This Chapter commences with a discussion of the role of meetings as a diffuser of new ideas and a catalyzer for knowledge creation. We also discuss how a firm utilizes meetings in knowledge production. In Section 14.3, we describe the characteristics of a spatial network of knowledge. Knowledge (or a significant part of it) is made available to firms by way of exchange processes in a spatial network of knowledge. In Section 14.4, we derive a production function for the knowledge levels of a nodal firm and the firm's accessibility measure to meetings by means of various networks of communication and exchange. A spatial equilibrium model of knowledge production over a network is formulated and the resulting Nash equilibrium is derived. In Section 14.5, a simple numerical example is used to illustrate that the system may bear endogenously the agglomeration cores in the distribution of knowledge production. It may also possess multiple equilibrium states depending on the parametric combinations of the relative nodal capacity of the meeting-facility and the technology adopted in knowledge production.

14.2 THE ROLE OF MEETINGS IN KNOWLEDGE PRODUCTION

14.2.1 Knowledge Production and Knowledge Exchange

The term, 'information', encompasses a broad spectrum of meanings. There has been an acceleration of studies of knowledge and information as instruments capable of improving the quality of life and products. The conceptual foundations of these studies are generally far too imprecise for a rigorous analysis of knowledge policies for the future (Andersson, 1985). We have to redefine certain concepts for our analysis if the ready-made meanings are not suitable. Our definitions of 'knowledge' and 'information' seem to possess narrower meanings than those which have been established in other fields of discourse. The problems arising from our needs have to be taken up for careful scrutiny. In this Chapter, however, we only propose a space-saving terminological agreement. The issues about terminology are open for further investigation.

According to our definitions, information implies a set of uncomplicated messages and routinized data which, to a large extent, can be subdivided into small pieces and stored with some simplicity for a longer or shorter time. The spread of telecommunications has increased the accessibility to information. Knowledge is intrinsically indivisible in nature. The transmission of knowledge requires face-to-face contacts among people, which ought to be diversified and even to some extent random. Knowledge exchange requires an extensive amount of somewhat diffuse movements throughout a transportation network. Knowledge production activities demand a high degree of accessibility to other persons.

So far, we have distinguished knowledge from information. We also have to differentiate between 'knowledge production' and 'knowledge exchange'. According to Machlup, 'knowledge production' designates any activity through which someone in a firm or an organization learns of something he or she had not known before, even if others knew about it (Machlup, 1980). Knowledge production also can involve both the creation of new knowledge and the search for new understanding from old knowledge. The term 'knowledge exchange' means any action which can contribute to the process of the disclosure, dissemination, transmission, and communication of knowledge. Knowledge exchange is essentially an interpersonal activity which should involve at least two persons. Knowledge production implicitly premises the exchange of knowledge among persons. In a broader sense, 'knowledge exchange' could become part of 'knowledge production'. The dichotomy between 'knowledge exchange' and 'knowledge production' can allow us to differentiate between the intra- and inter-oriented actions of firms. Thus, the term 'knowledge exchange' alludes to certain social phenomena behind the economic behaviour associated with a firm's knowledge production. These phenomena form the external economy of knowledge production in modern cities.

If one has a particular piece of knowledge and nobody else has it, knowledge exchange is the basic means to share it with all. Many different types of activities such as talking, listening, showing, debating, storming, intuiting, etc. can be employed simultaneously in some haphazard way to convey one's knowledge and the underlying concepts to others. The amalgamation of different concepts often results in the formation of something new. Such a creative feature of the process of knowledge exchange can be manifested in a form of dynamic synergy.

Independent contact between two persons is a basic vehicle for knowledge exchange. However, such independent communication is not always efficient. If all knowledge was exchanged on a bilateral basis, the number of contacts would easily go beyond the network capacity of the cities. Let us now imagine a situation in which five persons intend to exchange their knowledge to one another. If knowledge exchange is performed on a one-to-one basis, everybody is forced to have more than four bilateral contacts with the others. However, if the persons come together at the same time and place, they could intensively exchange their knowledge with one another, even in a limited time period. Thus, multi-person meetings could be efficient devices to save the potential frequency of individual contacts.

The other essential role of multi-person meetings is to provide a vehicle for participants to get to know those who know what they hope to know. It is difficult to learn what others know without any exchange of knowledge. It is most unlikely to become acquainted with those who possess the very knowledge that an individual hopes to gain through only randomly diversified bilateral contacts. Multi-person meetings can avoid potentially fruitless contacts in the search to become acquainted with the person possessing the relevant knowledge. As the scale of a meeting grows, the participants can expect to gain more knowledge about 'know-who'.

Highly knowledge-oriented firms tend to locate in metropolitan regions, seeking economies of agglomeration. Firms may choose to undertake their own R & D activity to expand their own stocks of knowledge, but eventually such knowledge stocks become available to other firms if the knowledge can be made accessible through the process of knowledge exchange. Thus, social interactions among firms, organizations, and individuals which facilitate any kind of diffusion of knowledge, form the external economy for knowledge production in metropolitan regions.

14.2.2 Knowledge as a Product

Knowledge can be classified either as an intermediate product required in the production of other goods and services or as a final product transacted in a market. 'Knowledge as an intermediate product' shares a significant role with other physical inputs in the production process of goods, services, and knowledge. Knowledge of this type is essentially paid for by others than those who use the knowledge. In other words, knowledge possesses no market price, and its price is absorbed into the current costs of some final product. 'Knowledge as a final product' ordinarily acquires a price in a knowledge market. This type of product can be subdivided into two sub-classes: (i) 'specified-knowledge products' and (ii) 'copied-knowledge products'. The former may take the form of tangible goods or intangible services, being produced in order to suit specified customers' requirements. The outcome is only available to a selected few. The content and quality of knowledge differ significantly over every product. The copied-knowledge products often take the form of corporeal goods. Knowledge may be written, printed, drawn, or engraved on paper, or encoded on disks, tapes, or other materials for the customers to read, listen to, and decode in order to understand it. The copied-knowledge products can be available, in principle, to all interested persons. The anonymity of customers is a premise in the transaction of products. Knowledge included in the same kind of product is the same, per se.

Meetings are organized by persons who are motivated to exchange their knowledge with one another. There should be some prototypes of meetings. 'Transaction-oriented meetings' such as negotiations, commissions, committees, etc., are organized by those directly or indirectly concerned with the transaction of knowledge. The meeting is usually oriented to produce something tentative, e.g., to make a tentative agreement on a transaction, to produce a preliminary outcome, to obtain the knowledge consumed in the current production, etc. 'Non-transaction-oriented meetings' typically take the form of conferences and conventions. Participants are encouraged to exchange a broader class of knowledge, which may bring reciprocated benefits in the long run, rather than enjoying current benefits in the short term.

The producers of specified-knowledge are obliged to have a series of transaction-oriented meetings with their customers, so as to clarify the clients' requirements. Knowledge workers of non-profit firms and governments are often invited to meetings, and sought to provide relevant knowledge for the production of outcomes. In contrast to this, the producers of copied-knowledge do not usually seek to know

Table 14.1 : The taxonomy of knowledge production firms in the Primary sector

	Specified-Knowledge Production Firms	Copied-Knowledge Production Firms	Non-profit Oriented Firms
Sub-sector related to the Market of Knowledge	research institutes in the private sector inspecting offices consultants law offices and lawyers business management services knowledge services countinghouses	newspapers (publications) journals and books (publications) movie producing and video services news syndicates information services education industries	universities research institutes schools
Sub-sector related to the Knowledge of Markets	insurance agencies publicity businesses advertisement agencies market research market information services brokerages	trade papers and journals (publications)	labour unions political powers

their customers in advance; they do not have direct contact with their customers. The customers are, in principle, precluded from meetings. Whether a meeting should be oriented to knowledge transaction or not is a rather arbitrary matter of judgment to some extent. The knowledge that is exchanged in meetings can be regarded as an intermediate product which is invested for knowledge production in the future or consumed in current production.

14.2.3 Knowledge Production Firms

M.U. Porat (1977) distinguishes two 'information sectors': a primary and a secondary one. The 'primary information' sector includes those firms which supply the bundle of information goods and services in a market context. Outcomes of the former possess prices and are transacted in a market. The latter includes all firms which produce information- and knowledge-services only for within-firm consumption. The primary sector can be subdivided into the 'sub-sector related to the market of knowledge' and the 'sub-sector related to the knowledge of markets'. Another taxonomy of the primary sector is also possible. We classify the knowledge production firms into three major subclasses as shown in Table 14.1: the 'specified-knowledge production firms', the 'copied-knowledge production firms', and the 'non-profit oriented firms'. This trichotomy shall open up vistas that might be relevant for our rigorous analysis of knowledge production, and will occupy us throughout our study.

The value to a customer of any knowledge of any tangible or intangible product is measured by what he would give in exchange for it. The distinguished trait of knowledge is that we can not know what a piece of knowledge may be worth to us before we know what it is. The value is based on expectation ex ante, not on regret ex post. It is determined by the satisfaction that the consumers anticipate. The anticipations by the customers are formed according to stacks of experiences from which they can learn and improve their judgments regarding valuations of knowledge that the producer has made. Knowledge included in each specified-knowledge product is not homogeneous per se. Every product is disposed of at a different valuation. The clients for the specified-knowledge products usually have quite a precise idea of how much they are willing to pay for it. Prices are decided by negotiations. On the other hand, knowledge included in the same kind of copied-knowledge products is intrinsically homogeneous. As no room for price negotiations is allowed, the market for the copied-knowledge may fall into the category of a quoted-price market, where the producers quote prices on a take-it-or-leave-it basis. Thus, the categories of knowledge markets and knowledge production firms should be manifold. No unified theory is likely to provide the common foundation to cope with all of the types of micro behaviour of knowledge production. Hereafter, our focus of this Chapter shall be on the micro behaviour of firms in the specified-knowledge production category.

14.3 MEETING-FACILITIES AND KNOWLEDGE ACCESSIBILITY

14.3.1 Knowledge Networks and Meeting-Facilities

Knowledge is made available to firms by way of exchange processes on a spatial network of knowledge. The term 'spatial network' is used to denote a set of nodes together with the links connecting the nodes. The nodes in a knowledge network can be characterized by their constellation of knowledge production capacities and pertinent activities; their knowledge infrastructure such as universities, meeting infrastructure, etc.; their stocks of knowledge and human capital; and their local networks of knowledge (Johansson, 1986). The links between nodes facilitate flows which comprise the displacement of messages, information and knowledge by making use of the dual network infrastructure, transportation and telecommunication networks. Knowledge production firms on the nodes may choose to undertake their own R & D activity to expand their own stocks of knowledge given the built environment and other infrastructure, but eventually such knowledge stocks become available to other units at other nodes on the knowledge network.

There should be interactions among the stocks of knowledge and their distribution over space in the form of knowledge production units located within different nodes of the knowledge network. A fundamental feature of knowledge production firms is the high frequency of communications among persons. If standardized information is exchanged, contacts can be substituted by telecommunications. However, when the knowledge to be exchanged has a high degree of indivisibility, face-to-face contacts requiring transportation are inevitable. The contact frequency can be reduced by multi-person meetings which enable intense multi-person contacts during limited time intervals.

The typical facilities where meetings take place in modern cities are offices, hotels, convention halls, and conference centres. These facilities can be characterized by their collective properties which are to be shared by users. The recent propensity towards office-automation in modern business firms is increasing the demand for office space in city centres. The excess demand for space derives from the rise in rents for office space in the CBD, and enforces many agents to disperse with the spacious meeting spaces within their offices. Thus, there are ever increasing needs for public meeting spaces which are open for many unspecified users.

The essential feature of a meeting-facility is that of a collective asset, whose utility is determined endogenously by its number of users. When somebody possessing relevant knowledge enters a set of members at a certain meeting, the availability of knowledge in the meeting automatically increases with positive benefits. However, the congestion of the meeting-facility may decrease the utility of the meeting. Spacious rooms for a meeting can contribute to the creation of the utility of the meeting. Thus, the provision of meeting-facilities is not only expected to catalyze and induce knowledge exchange among persons, but to enhance the utility of the meeting.

14.3.2 Knowledge Accessibility

A description of interdependencies in a spatial public good analysis is most conveniently handled with an accessibility representation. Accessibility measures can be regarded as the spatial counterparts of discounting. Thus they represent the distribution of public goods in a simple way that imposes a very clear structure upon the relationship between activities and their environment. The scale of a meeting, which is also an input for all members of the meeting, increases with increments in the number of members attending the meeting. On the other hand, another utility of public goods, the 'availability of spaces in meeting-facilities' decreases as the number of members attending the meeting increases. Thus, the service levels of our public goods, the meetings, are endogenously determined within a spatial network.

Let us define the scale of meetings of zone $j (j = 1, \cdots, n)$ by

$$SCALE_j = a_1(D_j)^{a_2}, \tag{1}$$

where $a_1, a_2 (a_1 > 0, a_2 > 0)$ are parameters. D_j is the number of members attending meetings of node j, whose values are endogenously determined within the system. The level of congestion in the meeting-facility can be described by

$$DEN_j = \frac{a_3 D_j}{Q_j}, \tag{2}$$

where, a_3 is a parameter, and Q_j is the capacity of the meeting-facility in node j. The services of meetings are produced by

$$ATT_j = a_4(SCALE_j)^{a_5}(DEN_j)^{-a_6}. \tag{3}$$

By substituting (1) and (2) into (3), we get

$$ATT_j = \alpha_1 Q_j^{\alpha_2} D_j^{\alpha_3}, \tag{4}$$

where $\alpha_1 = a_1^{a_5} a_3^{a_6} a_4$, $\alpha_2 = a_6$, and $\alpha_3 = a_2 a_5 - a_6$. Let us suppose $\alpha_1 > 0$, $\alpha_2 > 0$, and $\alpha_3 > 0$. We have already described the positive benefits of agglomeration within node j by (1), and the negative effects of congestion by (2) in the service production of meetings. Now the accessibility measure of node k to meetings of node j is given by

$$ACC_{kj} = \exp(-\beta d_{kj}) ATT_j, \tag{5}$$

where d_{kj} is an internodal distance between nodes k and j, and β is a parameter of friction.

Let us measure the scale of meetings by the number of participants, which can be depicted by the gravity model:

$$G_{kj} = \alpha N_k^{\gamma} f_{kj} Q_j^{\alpha_2} D_j^{\alpha_3}, \tag{6}$$

where $f_{kj} = \exp(-\beta d_{kj})$, N_k : the number of knowledge-handling workers of node k. By summing up (6) with regard to k, we have

$$D_j = \alpha_1 Q_j^{\alpha_2} D_j^{\alpha_3} (\sum_k N_k^{\gamma} f_{kj}). \tag{7}$$

Then, the scale of meetings of node j should be

$$D_j = \{\alpha_1 Q_j^{\alpha_2} (\sum_k N_k^{\gamma} F_{kj})\}^{\varepsilon}, \tag{8}$$

where $\varepsilon = 1/(1 - \alpha_3)$. The accessibility of node i to all meetings over the whole network is summarized as

$$ACC_i = \sum_j \{\phi_{ij} Q_j^{\delta} (\sum_k N_k^{\gamma} f_{kj})^{\sigma}\} \tag{9}$$

where $\phi_{ij} = \alpha_1^{\varepsilon} f_{ij}, \delta = \alpha_2 \alpha_3 \varepsilon$, and $\sigma = \alpha_3 \varepsilon$. Hereafter, let us call the L.H.S. of (9) 'knowledge accessibility'. By summing up both sides of (6) in terms of j, we can get the measure to explain how many times the k-th firm's workers attend meetings within the period of interest, which is given by

$$G_i = N_i^{\gamma} \sum_j \{\phi_{ij} Q_j^{\delta} (\sum_k N_k^{\gamma} f_{kj})^{\sigma}\}. \tag{10}$$

14.4 MICRO ECONOMIC ANALYSIS OF KNOWLEDGE PRODUCTION

14.4.1 Knowledge Markets and Knowledge Demand Functions

There are several insurmountable obstacles to be faced in approaching the micro behaviour of knowledge production. Most of the difficulties stem from the fact that there is no possible measure of outcome which would be logically separable from a measure of input. A surrogate for the standard measurement is required to cope with them. Although still far from a comprehensive theory, our tenet is that some of the intrinsic ways in which knowledge differs from other goods are explicitly taken into account in micro economic analyses of knowledge production.

The first feature of knowledge is that it should be represented by some means, e.g. tapes, prints, books, etc., if it is to be transacted. Most services of knowledge can not be separated from products. Hence, we lack a valuation of knowledge itself; in most industries, the consumer puts a value on the product by paying a price for it. The cost of reproduction is rather costless. Once knowledge is transacted in the market, the producer loses the ability to observe the consumers' re-transaction and reproduction of knowledge except in those cases where knowledge is protected by the rights reserved.

Next, we can point out that the quality as well as the quantity of knowledge are essential in the valuations of products. Some of the most important ways of producing knowledge require as inputs intermediate products largely made with inputs of labour from non-knowledge-producing occupations. Thus, the actions of 'creating new knowledge' and 'representing knowledge by some means', must be joined together in the production process. In contrast to most other goods, one's own knowledge will not as a rule be reduced after the transaction of knowledge to others. Knowledge can contribute in the production process without being used up. This implies that knowledge can share the role as an important input to the production process with other material capitals.

Finally, we must refer to the property that all (specified-knowledge) products are essentially differentiated in nature, even if they are made in the same firm. This fact leads us to the economic theory that the quantity of the firm's outputs should be measured according to how many differentiated products are produced rather than how many copies are reproduced. It must be noted here that our focus has been on specified-knowledge products, and that discussions about copied-knowledge products, which should rely upon other analytical strands, are ruled out hereafter.

So far we have discussed the differentiated properties of specified-knowledge products, but we have not referred to other dimensions related to interdependences between products. If the number of rival firms producing the differentiated products is small, oligopolistic interdependences emerge. For our purpose, we assume that the number of firms is large and that cross-elasticities of product demands are negligible. The decisions of the producers have to be guided by their expectations of their rivals' actions, but not by the firm's recognition of the impacts of their own actions on their rivals' actions. The economic picture we have mentioned briefly here is closely akin to that of the theory of monopolistic competition developed by Robinson (1933) and Chamberlin (1933). Our main concern here is not to rework this established theory, but simply to manifest our knowledge market in spatial dimensions.

In our spatial product markets, the customers on the network spend their money on knowledge offered by a number of firms, which are also distributed over a spatial network of knowledge. For simplicity, let us assume that each node of the network represents one firm among others. Firms on the network are competing with each other for a given level of the total demand for knowledge. The costs for transactions and deliveries of knowledge are assumed to be uniform over the whole network. Let us here denote as p_i the product price the i-th firm quotes, and X_i is the demand for the product of the i-th firm. The demand for the products, X_i is governed either by the relative price or the relative quality level of knowledge reserved. Let us define the relative price of the i-th firm's product by (p_i/P), where P is the general price of the standard product. Let the relative quality level of knowledge of the i-th firm be denoted by (v_i/V), which represents all of the shifts in demand which occur, depending upon the deviations in the quality level of knowledge from the standard one, V, in the industry. An individual firm can control the nominal value of products, p_i and the knowledge quality level of products v_i. The demand for the products of the i-th firm can be described by the following general demand function:

$$X_i = \bar{X} f(p_i/P, v_i/V) \tag{11}$$

where $\bar{X}$ is a parameter representing the the total demand power for products of the industry, whose level is determined within a regional economy. By taking a first-order approximation to the general product demand function (10), we get the following constant-elasticity schedule:

$$\log X_i = \log \bar{X} - \eta \log(p_i/P) + \xi \log(v_i/V). \tag{12}$$

where, η and ξ are price- and quality-elasticity in the demand schedule respectively, whose values are assumed to be constants greater than unity. We can rewrite Equation (11) in the following form:

$$X_i = \bar{X}(p_i/P)^{-\eta}(v_i/V)^{\xi}. \tag{13}$$

This is a very simple expression of the demand schedule for products of the i-th firm, but it is capable of demonstrating our views on the properties of specified-knowledge products in our monopolistic competitive market.

In the short run, the relative price and the relative knowledge quality level govern the divisions of products among firms. The parameters $\bar{X}, P, V$ are rather long-term determinants and are assumed to be fixed in the short term. In the long run, however, the demands for each firm's products are regulated by the other firms' decisions about the prices and the quality levels of the products. As the society's demands for knowledge grow, the value of $\bar{X}$ will increase. The learning processes will shift the knowledge quality level of the industry upwards, resulting in increases of the value of the standard price level, P and that of the standard quality level, V. If some firms are remiss in investing in the stocks of knowledge, the relative quality levels of their products will fall in the long run. Thus, they will lose their share of the knowledge market.

14.4.2 Consumptive- and Creative-Production of Knowledge

As discussed so far, the 'creation of new knowledge' and the 'representation of knowledge' must coexist in knowledge production. The creation of new knowledge designates any kinds of activities that are oriented to enhance the knowledge quality level of products. This can be done by creating new knowledge and new meaning from old knowledge and by acquiring the relevant knowledge through knowledge exchange with others. Let us call such actions contributing to the creation of new knowledge 'knowledge-creative production'. On the other hand, the representation of knowledge designates those actions which contribute to the physical production of knowledge products. Some of the most important ways of giving knowledge an appropriate feature as a product require inputs of physical effort such as typing, writing, drawing, etc. These actions are not always oriented to producing new knowledge, but rather to consuming the knowledge which has been obtained through the current knowledge-creative actions and through experiences in the past. Let us term the actions accompanied by physical performances 'knowledge-consumptive production'.

Let us consider some firms facing demand schedules (13) and consider the situation where they stop allocating their knowledge resources into knowledge creative production. It is very likely that their relative levels of knowledge quality (v_i/V) shall decline in due course. Their demand schedules will shift downwards and eventually they will loose their market share. Consider another situation where some firms intend to increase their expenditure on knowledge creative production; the quantities of the outputs are supposed to be fixed to certain levels. Then, their demand schedules shift upwards if the standard level of knowledge quality V in the industry is kept constant. It must be born in mind, however, that if these firms contemplate allocating all their resources into knowledge creation, they can not produce any products; they will eventually gain no profit. Thus, our firm would face an irreconcilable situation, the so-called 'quality-quantity trade-off'. To formalize it, let us assume that knowledge-creative productivity is strongly separable from knowledge-consumptive production technology. Let us now introduce two classes of production functions; i.e, a 'knowledge-creative production function' and a 'knowledge-consumptive production function', and their associated divisions in the labour force: a 'knowledge-creative labour force' and a 'knowledge-consumptive labour force'. Knowledge-creative labour includes any type of pure 'brain workers' who are literally expected to create new knowledge. In contrast to this, knowledge-consumptive labour contains a broader class of workers who are concerned with physical performances. Besides the researchers, designers, artists, and planners, etc., quite naturally, the executives, secretaries, and all the intermediary transmitters of knowledge come into the knowledge-consumptive category. Labour divisions in knowledge production firms, however, are often equivocal. For example, knowledge-creative workers do not allocate all of their available time to creative activities. Our dichotomy is rather expedient. Labour divisions shall be discussed further in future studies.

Our firm is assumed to produce differentiated products with heterogeneous inputs of knowledge labour, knowledge resources and capital goods. Rigorously speaking, those inputs jointly contribute to the current rate of both qualitative and quantitative outputs in short-run knowledge production. In the long run, however, it seems to be reasonable to assume that the rate of quantitative outputs is mainly regulated by the input levels of material capital goods, the non-knowledge labour force, and information resources. The long-run quantitative technological relations between the rate of the final outputs and the required levels of inputs can be summarized by a knowledge-consumptive production function. It also appears that a large amount of capital goods is not required in knowledge-creative activities in our category of firms. The availability of knowledge in the node where firms locate is rather crucial in knowledge creation. Thus, we assume that the long-run relations between the rate of qualitative outputs and the input levels of the knowledge labour force can be manifested by a knowledge-creative production function. The availability of knowledge should also be included in this function as a variable.

We assume that the rate of quantitative outputs is regulated by a knowledge-consumptive production function which explains technological relations between the amounts of the final products X_i and the required level of inputs. This is described by a Cobb-Douglas production function:

$$X_i = b_1 M_i^{b_2} O_i^{b_3} D_i^{b_4} B_i^{b_5}, \tag{14}$$

where M_i : the number of knowledge-consumptive workers, O_i : the area of floor space, D_i : the capacity of information processing systems, B_i : the capacity of the data base available at node i. If a decreasing degree of returns to inputs is assumed, we get $b_2 + b_3 + b_4 + b_5 < 1$. If $M_i = 0$, we have $X_i = 0$ by (14). This concurs with the fact that our firm could gain nothing from the all-to-knowledge-creation policy as mentioned in the extreme case earlier.

A knowledge-creative production function explains the relations between the levels of inputs and the rate of qualitative output represented by a shift-parameter v_i of the demand schedule (13). Let us adopt the simplest specification of the knowledge-creative production function. It is assumed that the knowledge-creative production is regulated by two factors, i.e. the knowledge-creative labour force and the nodal accessibility to knowledge. Under these assumptions, our knowledge-creative production function can be formulated by

$$v_i = \mu_1 N_i^{\mu_2} ACC_i^{\mu_3}, \tag{15}$$

where N_i is the number of knowledge-creative workers, μ_1, μ_2 and μ_2 parameters.

14.4.3 Micro Behaviour of Knowledge Production Firms

Consider the inverse function of the demand schedule (13) which expresses the price of products as a function of the quantity demanded:

$$p_i = p(X_i, v_i) = \sigma_1 X_i^{-\sigma_2} v_i^{\sigma_3} \tag{16}$$

where $\sigma_1 = P\bar{X}^{\sigma_2} V^{-\sigma_3}$, $\sigma_2 = 1/\eta$, $\sigma_3 = \xi/\eta$. The problem of the (monopolistic competitive) firm facing the demand function (13) is then

$$\begin{aligned} & \max\{p(X_i, v_i)X_i - \omega_1 N_i - \omega_2 M_i - \omega_{3i} D_i - \omega_4 O_i - \omega_5 B_i\} \\ = \; & \max\{\sigma_1 X_i^{(1-\sigma_2)} v_i^{\sigma_3} - \omega_1 N_i - \omega_2 M_i - \omega_{3i} D_i - \omega_4 O_i - \omega_5 B_i\} \end{aligned} \tag{17}$$

where $X_i = b_1 M_i^{b_2} O_i^{b_3} D_i^{b_4} B_i^{b_5}$; $v_i = \mu_1 N_i^{\mu_2} (ACC_i)^{\mu_3}$; ω_1, ω_2: the wages of knowledge-creative and -consumptive labour, respectively; ω_{3i}: the rent of office floor space; ω_4, ω_5 : the rental costs per capacity of information processing systems and data base systems, respectively. Assume that the factor prices, except the rental of office floor space, are same in every node. Our firms choose the optimal levels of inputs and outputs so as to maximize their profits.

The producer's cost function C is defined as the solution to the problem of minimizing the cost of producing at least output level $\bar{Q}_i$. Assume here for the moment that the knowledge quality level and the number of knowledge-creative workers are fixed to the following levels, $\bar{v}_i$ and $\bar{N}_i$, respectively. Then our problem is

$$\begin{aligned} & \min_{M_i, D_i, O_i, B_i}\{\omega_1 \bar{N}_i + \omega_2 M_i + \omega_{3i} D_i + \omega_4 O_i + \omega_5 B_i\} \\ & \text{subject to} \\ & \bar{Q}_i = \sigma(b_1 M_i^{b_2} O_i^{b_3} D_i^{b_4} B_i^{b_5})^{1-\sigma_2} \bar{v}_i^{\sigma_3}. \end{aligned} \tag{18}$$

Then, the cost function is

$$C(\bar{Q}_i) = \kappa\pi^{-1/\kappa} \bar{Q}_i^{1/\kappa} \bar{v}_i^{-\sigma_3/\kappa} - \omega_1 \bar{N}_i, \tag{19}$$

where $\kappa = (b_2+b_3+b_4+b_5)(1-\sigma_2)$, $\pi_i = \sigma_1(1-\sigma_2)^{\kappa} b_1(b_2/\omega_2)^{b_2}(b_3/\omega_{3i})^{b_3}(b_4/\omega_4)^{b_4}(b_5/\omega_5)^{b_5}\}^{(1-\sigma_2)}$. Substituting (19) into (17), we get

$$\max_{Q_i}\{Q_i - \kappa\pi_i^{-1/\kappa}Q_i^{1/\kappa}\bar{v}_i^{-\sigma_3/\kappa} - \omega_1\bar{N}_i\}. \quad (20)$$

Based upon the optimal solution of (20), the optimal supply function Q_i^* can be constructed as:

$$Q_i^* = \psi_i\bar{v}_i^{\sigma_3/(1-\kappa)}, \quad (21)$$

where $\psi_i = \pi_i^{1/(1-\kappa)}$. Assume that $b_2+b_3+b_4+b_5 < 1$. Then we can define the profit function which provides yet another way in which a decreasing returns to scale technology can be described:

$$\Pi(\bar{N}_i) = \{(1-\kappa)\psi_i\bar{v}_i^{\sigma_3/(1-\kappa)} - \omega_1\bar{N}_i\}. \quad (22)$$

Let us derive the factor demand functions for knowledge resources by applying Hotelling's lemma. By differentiating the profit function with respect to each of the factor prices respectively, we get

$$\begin{aligned} M_i &= (b_2/\omega_2)(1-\sigma_2)\pi_i^{1/(1-\kappa)}\bar{v}_i^{\sigma_3/(1-\kappa)}, \\ O_i &= (b_3/\omega_{3i})(1-\sigma_2)\pi_i^{1/(1-\kappa)}\bar{v}_i^{\sigma_3/(1-\kappa)}, \\ D_i &= (b_4/\omega_4)(1-\sigma_2)\pi_i^{1/(1-\kappa)}\bar{v}_i^{\sigma_3/(1-\kappa)}, \\ B_i &= (b_5/\omega_5)(1-\sigma_2)\pi_i^{1/(1-\kappa)}\bar{v}_i^{\sigma_3/(1-\kappa)}. \end{aligned} \quad (23)$$

Now, let us consider the problem of finding N to maximize the profit function (22). Substituting (15) into (22), we get

$$\max_{N_i}\{\Phi_i[N_i^{\lambda_1}(ACC_i)^{\lambda_2}] - \omega_1 N_i\}, \quad (24)$$

where $\lambda_1 = \mu_2\sigma_3/(1-\kappa)$, $\lambda_2 = \mu_3\sigma_3/(1-\kappa)$, $\Phi_i = \mu_1(1-\kappa)\psi_i$. Given the decisions of the firms other than the i-th, let us maximize (24) with respect to N. Let us substitute (9) into (24), and derive the first-order optimal conditions of the problem (24). After a little rearrangement of the optimality conditions, we get

$$\begin{aligned} N_i \;=\; & (\Phi_i/\omega_1)\{\lambda_1 N_i^{\lambda_1}(ACC_i)^{\lambda_2} \\ & +\lambda_2 N_i^{\lambda_1+1}\cdot(ACC_i)^{\lambda_2-1}\partial(ACC_i)/\partial N_{i(sub)}\}, \end{aligned} \quad (25)$$

where $\partial(ACC_i)/\partial N_{i(sub)}$ represents the subjective expectation of marginal accessibility with respect to the nodal knowledge-handling labour force. Suppose the nodal firms have myopic expectations on marginal accessibility, i.e., $\partial(ACC_i)/\partial N_{i(sub)} = 0$. Then, the optimal condition can be simplified to

$$N_i = \{(\Phi_i/\omega_i)\lambda_1(ACC_i)^{\lambda_2}\}^{1/(1-\lambda_1)}. \tag{26}$$

Represent the R.H.S. of (26) by $F_i(N)$; then (26) becomes

$$N_i = F_i(N). \tag{27}$$

Since we assume for the moment that $N_j (j \neq i)$ are fixed to certain levels, Equation (27) becomes a fixed point problem with respect to N. Let N be a fixed point of (27). By substituting N into (23), the factor demand profiles of knowledge resources can also be derived.

14.5 SPATIAL EQUILIBRIUM OF KNOWLEDGE PRODUCTION

14.5.1 Spatial Nash-Equilibrium of Firms

The theory of urban economics is, to a large extent, based on the principles outlined in seminal works by von Thünen. The basic exogenous factors of location theories in the school of urban economics were a pre-designed transportation network, a form of technology for using this network and a pre-located concentration of work and service activities. The basic von Thünen assumption of a central market place, later transformed into the existence of a central business district, was also a basic external assumption in new urban economics.

Our contributions build on a development of spatial equilibrium theory that was initiated by Karlqvist and Lundqvist (1972), Andersson and Karlqvist (1976), etc. The basic difference between new urban economics and new-new urban economics in the Lundqvist-Andersson-Karlqvist approach was to discard the assumption of some pre-located market place or central business district. Their basic assumption was to include the location of activities through a deeper form of interdependency analysis, where only transportation networks and technology were used as exogenous factors of location (Andersson and Ferraro, 1983; Andersson and Kobayashi 1989). Contributions by Batten, Kobayashi and Andersson (1989) can also be related to this development.

The basic assumption of this new approach is the principle of interdependencies between decision makers. It comes close to the basic principle of game theories, which explicitly assume that the decision of a given decision maker is contingent upon simultaneous decisions by other decision makers. It is also an approach in some harmony with arguments from 'agglomeration economics'.

Let us imagine a situation where meeting-facilities have already been allocated to nodes as local public goods. Every firm is supposed to be free to choose an optimal level of output, given the current values of exogenous variables. In our knowledge production distribution, some of the attractiveness of nodes, i.e. the accessibility to meetings, is influenced by the spatial distribution of the knowledge production of other nodes. If some of the parameters of the system affecting the accessibility change, it is likely that the spatial equilibrium will also change. The resulting change in knowledge distribution determines a new level of the attractiveness of each node, which in turn affects its relative attractiveness. If the knowledge distribution is in disequilibrium as a result of changes in the exogenous variables, sooner or later each firm modifies its current decisions until a new equilibrium state is reached; i.e., the level of knowledge accessibility and the spatial distribution of knowledge production equilibrate.

The spatial equilibrium derived by non-cooperatively competing firms falls within the category of a Nash-equilibrium. The criterion generating the equilibrium solution is that the optimality conditions for knowledge production levels should be satisfied simultaneously in every location. This can be induced from the description of a possible pattern of reactions of agents, free to choose their optimal levels of output in a non-cooperative way. Thus, our equilibrium model can be given by a solution of the following fixed point problem:

$$N_i = F_i(N), (i = 1, \cdots, n). \tag{28}$$

14.5.2 Existence of Spatial Equilibrium

The vector function F of Equation (28) is defined on $C = (0, \infty]$. Since the set C is not compact, the ordinary fixed point theorem, say Brouwer's fixed point theorem (see, for example, Border, 1985), can not be applied to ascertain whether there exist fixed points in (28). In what follows, we try to derive the necessary conditions which guarantee the existence of fixed points in (28). Let us assume that the wage levels of knowledge-creative labour, ω_1, are to be endogenously determined to equilibrate the sum of N_i^* in terms of i, whose values are determined by solving the fixed point problem (28) and the total number of available workers in the region, N.

Assume the existence of fixed points in (28) given the value of $\bar{\omega}_1$. Denote a spatial equilibrium by $N_i^*(\bar{\omega}_1), (i = 1, \cdots, n)$ and the total number of knowledge-creative workers by $\bar{N}$, which imposes the adding-up constraint in spatial divisions of knowledge production. Assume that there exists a equilibrium wage ω_1^*, satisfying

$$\bar{N} = \sum_i N_i^*(\omega_1^*). \tag{29}$$

Let $\theta_i^* = (\Phi_i/\omega_1^*)^{-b_3(1-\sigma_2)/(1-\kappa)}$. Then, for an equilibrium state (N^*, ω_1^*), there holds

$$N_i^* = \{\theta_i^* \lambda_1 (ACC_i)^{\lambda_2}\}^{1/(1-\lambda_1)}. \tag{30}$$

By summing up both sides of Equation (30) with respect to i,

$$\begin{aligned} \bar{N} &= \sum_i N_i^*(\omega_1^*) \\ &= \sum_i \{\theta_i^* \lambda_1 (ACC_i)^{\lambda_2}\}^{1/(1-\lambda_1)} \end{aligned} \tag{31}$$

holds for the equilibrium state (N^*, ω_1^*). Then, N_i^* satisfies

$$N_i = \frac{\bar{N}\{\theta_i^*(ACC_i)^{\lambda_2}\}^{1/(1-\lambda_1)}}{\sum_k \{\theta_k^*(ACC_k)^{\lambda_2}\}^{1/(1-\lambda_1)}}. \tag{32}$$

To the contrary, let us assume that θ_i^* in (32) is fixed to a certain value $\theta_i^{(0)}$. Since the map (32) is continuous on the compact set $\Delta = \{N_i : \sum_i N_i = \bar{N}\}$, the existence of fixed points in (32) is guaranteed by Brouwer's fixed point theorem. Denote a fixed point of Equation (32) by $N(\theta^{(0)})$ for an arbitrary $\theta^{(0)}$. If there exists a θ^* satisfying (31), $N(\theta^*)$ obviously satisfies (31). In other words, if there exists a θ^*, satisfying (32) for a predetermined level of $\bar{N}$, there exists an equilibrium wage ω_1^* in (29). Thus, the existence of θ^* satisfying (31) is necessary for the existence of an equilibrium wage in (29). Since θ_i is a uni-parameter function of ω_1, the existence of fixed points in (28) can easily be ascertained by applying the standard algorithms for solving fixed point problems.

14.5.3 Numerical Illustration

There are enormous complexities beneath the apparent simplicity of the representation of our spatial equilibrium model (32). One of the most difficult dynamic phenomena to model theoretically is the discontinuous movement of spatial equilibrium states occurring as some parameters move smoothly into critical regions that mark regime changes in qualitative features of spatial equilibriums. Fixed points in (32) can change their nature at critical values of parameters and the system can jump between alternative equilibrium states as a result of perturbations. 'Historical accidents' can have a crucial impact on the spatial form of the distribution of knowledge production. The nature of transformations between equilibrium states can be investigated using numerical experiments in relation to the following idealized system.

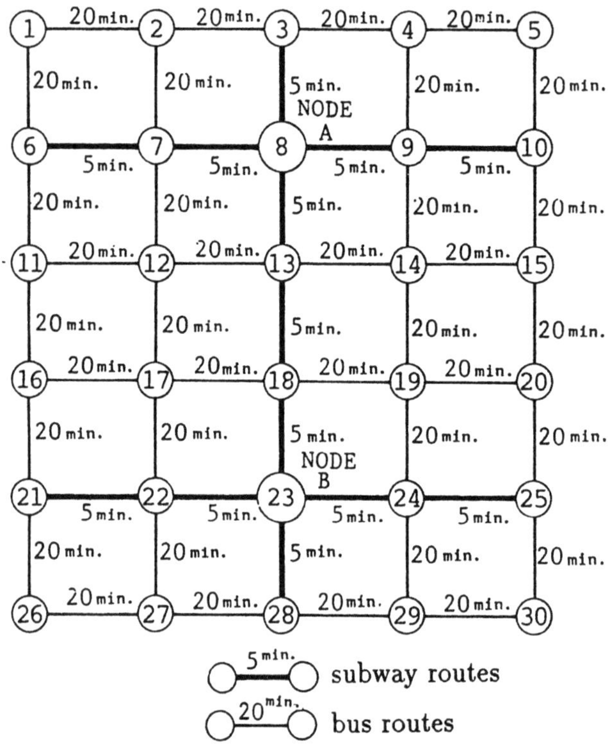

Numerals show internordal time distances.

Figure 14.1 : A hypothetical discrete network

Consider a hypothetical city described by a discrete network as shown in Figure 14.1. There are thirty nodes, each of which represents a centroid of each zone. Assume for simplicity that the land rent of each node takes the same value. 150 knowledge-creative workers are to be allocated in the city and of obvious interest is the spatial equilibrium of knowledge-creative workers over the discrete network. The data for the network are also described in Figure 14.1. Both nodes A and B represent the major junctions of the subways. For the moment, let us assume that meeting-facilities are allocated uniformly over the network; the capacity of the meeting-facility for each node is supposed to take the same value, i.e. $Q_j = 2.0 (j = 1, \cdots, 30)$.

Figures 14.2 and 14.3 illustrate what will happen to spatial divisions of knowledge-creative workers in two cases where the values of parameters are fixed to the following, respectively: (i) $\beta = 0.01, \gamma = 2.0, \delta = 2.0, \lambda_3 = 2.0$, (Case 1); (ii) $\beta = 0.05, \gamma = 2.0, \delta = 2.0, \lambda_3 = 2.0$, (Case 2). In both cases, it is obvious that transport systems have a

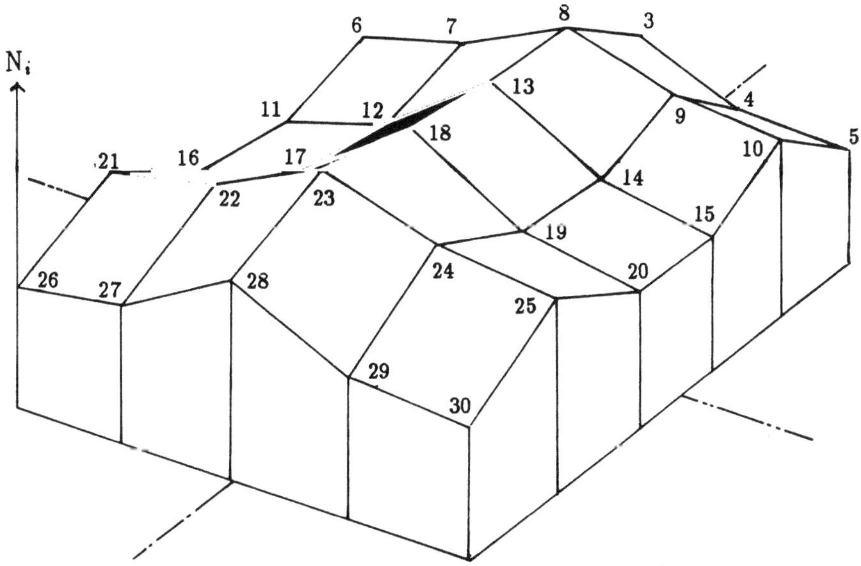

Figure 14.2 : A spatial division of the knowledge-creative labour force (where $\beta = 0.01$, $\gamma = 2.0$, $\delta = 2.0$, $\lambda_3 = 2.0$) (Case 1)

significant influence on spatial divisions of workers; our system bears agglomeration cores of knowledge production endogenously within the discrete network. Different qualitative regimes of spatial equilibriums can be clearly identified in Figures 14.2 and 14.3. Case 1 makes a bicipital distribution of knowledge production, which is bisymmetrical about the horizontal and vertical centre lines. Case 2 apparently produces two different asymmetrical patterns about the horizontal central line. Initial configurations of knowledge production determine the qualitative regimes of stable equilibriums.

We have already seen in Figures 14.2 and 14.3 that transport systems have major impacts on the formation of the urban structure. It is possible, and interesting, to explore such impacts more systematically. Examples of possible states for spatial equilibriums of knowledge production as a function of two parameters can also be studied (γ is associated with the scale economy in knowledge exchanges, β with ease of travel). The larger γ, the more important is synergy in knowledge

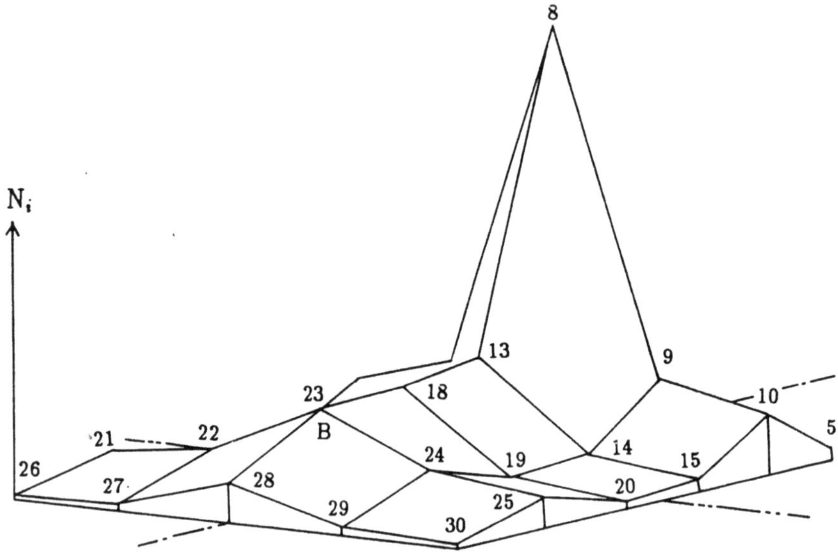

a) Spatial Equilibrium (Type 1)

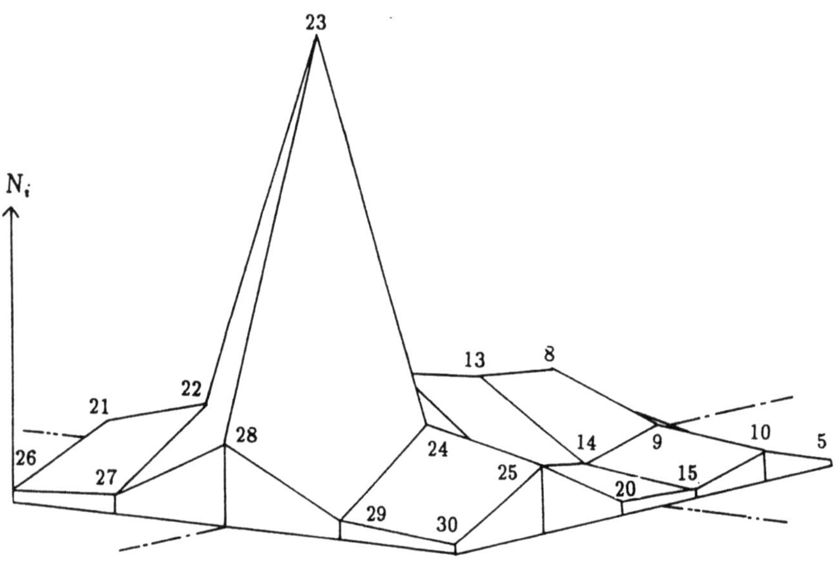

b) Spatial Equilibrium (Type 2)

Figure 14.4 : Different types of spatial equilibrium for a knowledge-creative labour force (where $\beta = 0.05$, $\gamma = 2.0, \delta = 2.0, \lambda_3 = 2.0$) (Case 2)

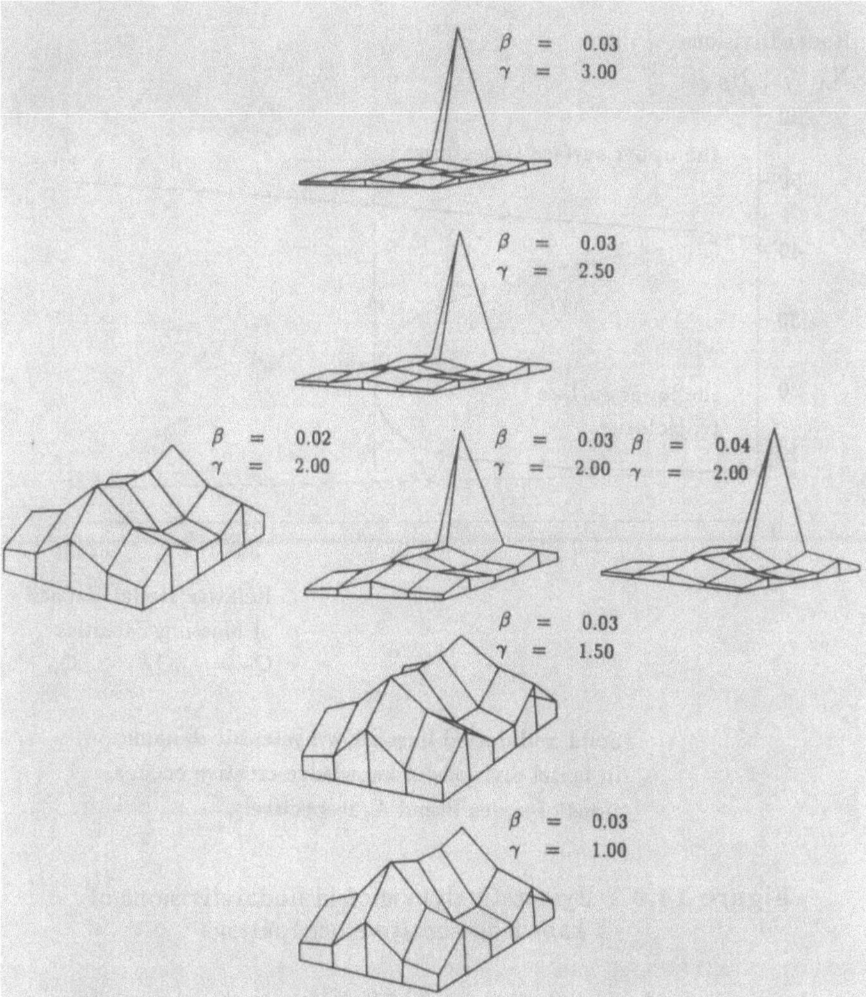

Figure 14.5 : Regime transitions of spatial equilibrium (where $\delta = 2.0, \lambda_3 = 2.0$)

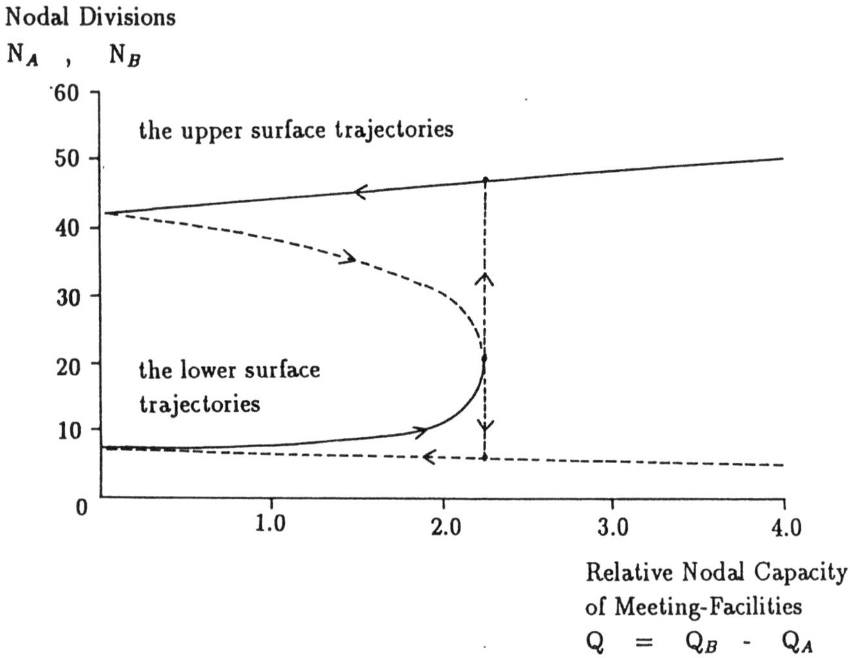

Solid and dotted lines show hysteretic dynamics in nodal divisions of knowledge-creative occupations of nodes B and A, respectively.

Figure 14.6 : Hysteretic dynamics in nodal divisions of knowledge-creative occupations

exchanges; the larger β, the more difficult in general travel is. Let us consider the initial configurations where all knowledge production firms are concentrated in node A. Computer simulations are carried out for each combination of parameters γ and β. Figure 14.4 explains transformations between stable equilibrium states. As γ and β become large beyond certain critical values, there emerge asymmetric patterns of knowledge equilibriums; the system possesses multi equilibrium states.

Let us now investigate precisely the qualitative changes of spatial forms in terms of the capacity of a meeting-facility. Figure 14.5 illustrates how qualitative features of stable equilibrium states change if the relative nodal capacity of meeting facility, $Q = Q_B - Q_A$, changes. Q_A and Q_B represent the capacity of meeting-facilities in nodes A and B, respectively. The pivotal case in Figure 14.4, where $\beta = 0.03, \gamma = 2.0, \delta = 2.0, \lambda_3 = 2.0$, is regarded as the initial state in this experiment. In Figure 14.5, N_B represents the number of knowledge-creative workers in node B. Let us assume that Q is initially about 0.0, corresponding to an equal proportion of the capacity of meeting-facilities between nodes A and B. If the capacity of the meeting facility in node B grows (for example, due to the opening of new convention halls or conference centres), the value of N_B rises rather slowly until the critical point is reached. There, N_B jumps up to the higher surface and the relative allocation of knowledge-creative workers changes quite abruptly. A key feature of this type of nonlinear dynamics is its cyclic nature. If the nodal capacity of the meeting-facilities in node B deteriorates, the system may follow the upper-surface trajectory (with N_B decreasing slowly) until a second critical point is encountered. Here N_B falls to the lower surface and node B's share of the knowledge-creative workers drops suddenly. This type of hysteresis loop is useful as a tool for studying qualitative changes in the creative capacity of regions.

14.6 CONCLUSION

This Chapter has shown that location theories of knowledge production should be based on a foundation other than the traditional paradigm of location in relation to some predetermined market place or CBD. We have argued that the formation of such centres should be endogenously produced given some preconditions about urban infrastructure. This new paradigm of location should be sought within a theory of interdependent networks. The Chapter has introduced some ideas towards a more comprehensive theory of interdependent networks of knowledge production.

It must be stressed that this study is not by any means a complete study of the interdependencies of knowledge production. Our main conclusion is that the study can serve as a foundation for the development of quantitative and computationally-oriented equilibrium approaches to location theory. It is also possible to use these approaches to evaluate possible changes in the distribution of knowledge production between nodes as a result of the construction of meeting-facilities and network infrastructure, as well as other major infrastructural changes in the region. Thus, our approaches are beneficial for planning debates about the role of knowledge infrastructures in the formation of creative regions.

REFERENCES

Andersson, Å.E., 1985, *Creativity: The Future of the Metropolis*, Prisma, Stockholm (in Swedish).

Andersson, Å.E. and A. Karlqvist, 1976, "Population and Capital in Geographical Space. The Problem of General Equilibrium Allocation", in Los, J. and W. Los (eds.), *Computing Equilibria: How and Why*, North-Holland, Amsterdam.

Andersson, Å.E. and G. Ferraro, 1983, "Accessibility and density distributions in metropolitan areas: Theory and empirical studies", *Papers of the Regional Science Association* 52:141-158.

Andersson, Å.E and K. Kobayashi, 1989, "Some Theoretical Aspects of Spatial Equilibria with Public Goods", in Å.E. Andersson, D.F. Batten, B. Johansson, and P. Nijkamp (eds.), *Advances in Spatial Theory and Dynamics*, North-Holland, Amsterdam, pp. 207-221.

Batten, D. F., Kobayashi, K. and Å. E. Andersson, 1989, "Knowledge, Nodes, and Networks, An Analytical Perspective", in Andersson, Å.E., Batten, D.F., and C. Karlsson (eds.), *Knowledge and Industrial Organization*, Springer-Verlag, Heidelberg, pp. 31-46.

Border, K.C., 1985, *Fixed Point Theorems with Applications to Economics and Game Theory*, Cambridge University Press, Cambridge.

Chamberlin, E.H., 1933, *The Theory of Monopolistic Competition*, Harvard University Press, Cambridge, MA.

Johansson, B., 1986, "Viability of spatial networks", paper presented at the RSA Conference in Columbus, Ohio.

Karlqvist, A. and L. Lundqvist, 1972, "A contract model for spatial allocation" *Regional Studies*, 4:401-19.

Machlup, F., 1980, *Knowledge and Knowledge Production*, Princeton University Press, Princeton.

Porat, M.U., 1977, *The Information Economy: Definition and Measurement, Vol.1*, U.S. Department of Commerce, Washington, DC.

Robinson, J., 1933, *The Economics of Imperfect Competition*, Macmillan, London.

CHAPTER 15

Dynamic Change of Urban Housing Stock, Construction and Demolition

Masuo Kashiwadani

15.1 INTRODUCTION

The population of most major cities in developed countries has not increased very much in the past decade. Due to this phenomenon, these cities could be expected to become mature and stable. However, the stable condition of the population does not mean that cities have ceased to grow or change. Even those cities in which the population is declining enlarge their areas of development and change their land use patterns. The cities which we will study in developed countries do not grow rapidly, but their form changes intensively.

It is very difficult to project the future land use pattern or activity distribution in these cities. The most empirical urban land use models are made for growing cities. They are called incremental models. First we determine the whole increment of each activity in a city. Then we allocate a share of it over tracts in the city according to some location scores calculated in the model. Incremental models are not useful for those cities in which the whole volume is stable but the land use pattern is changing. Suppose a city where the volume of an activity is increasing in some tracts and decreasing in other tracts, while the volume of the activity for the city over all is quite stable. Thus, the increment would be almost zero and we could not allocate any volume increase or decrease.

One way to solve this problem is to use a stock allocation model instead of the incremental allocation model. But this would be contrary to the historical development of empirical models during recent decades. The forecasting fitness levels of incremental models have been much better than those of stock allocation models in the case of most growing cities (Putman, 1975). Another way to retain the advantage of incremental models is to separate the activity which should be allocated into two parts - the increasing part and the decreasing part - and to make two allocation models of the increment type for each part.

But it is very hard to project or estimate the number of houses demolished in a practical sense. Accessible data about demolition is rare. It is difficult to define housing demolition as a conceptually clear and practically useful term. Only two models, the NBER model (Ingram et al., 1972) and the Dortmund model (Wegener, 1982), have so far tried to deal with demolition. Both models are represented by a simulation model using quite specific data which is obtained by special investigations.

Usually empirical land use models are represented by equations with parameter values statistically estimated using data which is accessible to ordinary researchers. The final purpose of this study is to make a forecasting model of housing distribution in a metropolitan area which is separated into a construction submodel and a demolition submodel.

15.2 HOUSING CONSTRUCTION AND DEMOLITION IN OSAKA PREFECTURE

Our study area is Osaka Prefecture. The population of Osaka Prefecture is about 8.6 million and the area forms the major part of the Osaka Metropolitan Area of which the population is 11.8 million.

Table 15.1 shows the number of dwellings in the area between 1968 and 1988 by building type. Here the value of the housing stock is taken from the data of the Housing Survey of Japan which is undertaken every 5 years. The number of dwellings constructed is taken from the Annual Statistical Report of Buildings (*Kenchiku Chakko Tokei Nempo*). The numbers in the Table show those dwellings constructed in the period between two housing survey years. Actually they are the sum of the number of dwellings constructed from the year when the former housing survey is completed to the year before the next housing survey. This method is adopted to adjust the difference between the two sources of data. The number of dwellings demolished in a period is estimated by subtracting the number of dwellings in the latter housing survey from the sum of the number of dwellings in the former housing survey and the number of dwellings constructed during the two housing survey years.

Recently, the total housing stock has increased gradually while the population has been almost stable. The rate of increase of the total

Table 15.1 : The number of dwellings by building type in Osaka prefecture (thousands)

year / type	1968 Stock	C	D	1973 Stock	C	D	1978 Stock	C	D	1983 Stock	C	D	1988 Stock
wooden detached or tenement	1130.8	281.8	117.7	1294.9	234.8	174.8	1354.8	176.2	127.4	1403.6	109.4	120.3	1392.7
wooden apartment	446.8	121.0	47.2	520.6	20.5	71.7	469.4	3.0	97.2	375.2	4.4	45.2	334.4
non wooden detached or tenement	47.0	44.0	7.0	84.0	44.0	27.7	100.3	48.4	40.7	108.0	45.3	27.1	126.2
non wooden apartment	179.3	235.4	26.9	387.8	246.2	65.3	568.7	208.5	30.6	746.6	294.9	33.9	1007.6
total	1803.9	682.2	198.8	2287.3	545.5	339.6	2493.2	436.1	295.9	2633.4	454.0	226.5	2860.9
population	7218.9			8068.2			8473.5			8588.8			8559.0

C: construction
D: demolition

housing stock is less than 10% but the rate of dwellings constructed compared to that of the housing stock shows more than twice that rate of increase except for the period 1968-1973. This means that the number of dwellings demolished has not been small. It is about 10% of the total housing stock on the average. We can see that the pattern of construction and demolition is different for different building types; moderate construction and demolition has occurred in wooden detached or tenement houses; there has been a very low construction rate and much higher demolition rate of wooden apartments; there have been both high construction and demolition rates of non wooden detached or tenement houses; and a high construction rate and low demolition rate of non wooden apartments.

Table 15.2 shows the housing distribution in rings around Osaka Prefecture by building type. The number of detached houses in the prefecture is increasing but that of the central city ring is decreasing. As for tenement houses and low rise apartments, the total number of houses is decreasing and the number of houses in the rings is also decreasing, except in the outer suburbs. The number of middle rise or high rise apartments is increasing in every ring. In particular, the increase in the number of high rise apartments in the central city ring is larger than the increase in any other ring. The change in the urban

Table 15.2 : Housing distribution by ring and building type

		detached house	tene-ment house	low-rise apart-ment	mid.-rise apart-ment	high-rise apart-ment	total
Osaka city Central Area	1980(A)	32,937	14,757	12,005	22,771	21,892	104,362
	1985(B)	31,214	12,405	9,407	24,962	38,115	116,103
	(B)-(A)	-1,723	-2,352	-2,598	2,191	16,223	11,741
Osaka city peripheral	1980(A)	173,110	237,442	140,819	123,094	81,216	755,681
	1985(B)	185,183	205,855	131,811	147,636	127,890	798,375
	(B)-(A)	12,073	-31,587	-9,008	24,542	46,674	42,694
inner suburbs in Osaka prefecture	1980(A)	339,595	192,505	160,284	150,255	64,487	907,126
	1985(B)	372,251	173,417	146,482	169,455	93,795	955,400
	(B)-(A)	32,656	-19,088	-13,802	19,200	29,308	48,274
outer suburbs in Osaka prefecture	1980(A)	318,037	110,712	95,432	121,217	26,890	672,288
	1985(B)	471,456	114,320	97,110	147,515	41,618	872,019
	(B)-(A)	153,419	3,608	1,678	26,298	14,728	199,731
Osaka prefecture total	1980(A)	863,679	555,416	408,540	417,337	194,485	2,439,457
	1985(B)	1,060,104	505,997	384,810	489,568	301,418	2,741,897
	(B)-(A)	196,425	-49,419	-23,730	72,231	106,933	302,440

housing distribution pattern is so complex that we need to classify dwellings into several building types and to deal with construction and demolition separately when we forecast the future dwelling distribution in a city.

There are some significant reasons for the recent level of demolition of urban dwellings in Japanese cities. The first is demolition due to dilapidation. The second is demolition due to urban redevelopment. The third is demolition due to market selection. It is supposed that the first reason is related to the age of the house, the second to the location and the third to housing types respectively. Some investigative data have shown that the first and third reasons are the most important. However, the fact that older houses are demolished in greater numbers than new houses means that older houses have had much more time to be exposed to the chance of demolition. Another empirical study questions whether the demolition rate over a term may be unrelated to classification by the house's age (Kashiwadani and Kawachi, 1989). On the other hand the difference in the housing demolition rate among different housing types is quite distinctive, as shown in Tables 15.1 and 15.2. It is suggested by Table 15.1 that the number of dwellings demolished by housing type is affected by the number of dwellings constructed and the market response to the housing construction supply.

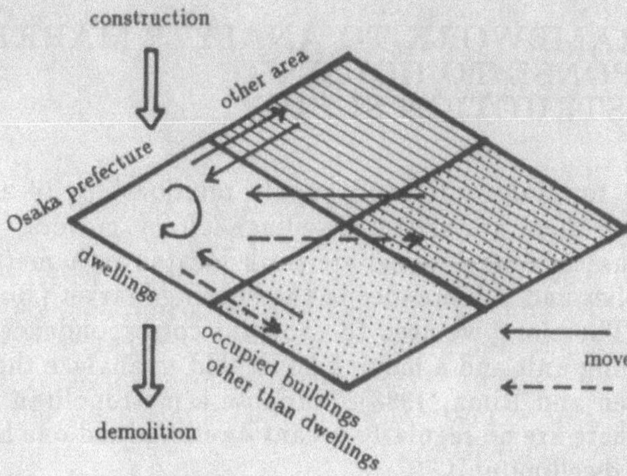

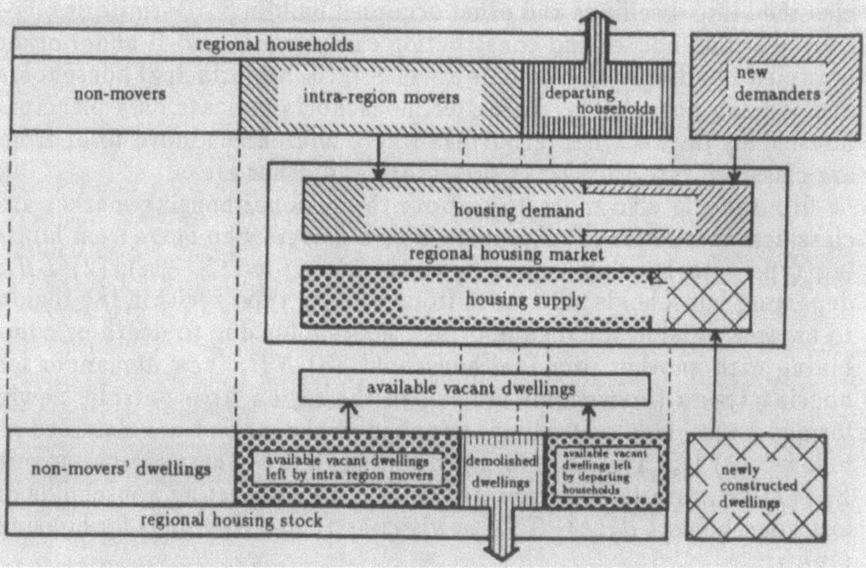

Figure 15.3 : **Changes of households and dwellings in a regional housing market**

15.3 A FRAMEWORK TO ANALYZE MARKET RESPONSE TO HOUSING CONSTRUCTION SUPPLY

It is not easy to analyze demand supply relationships in a housing market because there are vacant dwellings left by movers. A mover whose old house and new house are both located in a metropolitan area is a supplier and a demander in the housing market (Rothenberg et al., 1991). Therefore, we need to know the correspondence between a vacant dwelling unit and a moving household to analyze the housing market (Wissen and Rima, 1988). Suppose a metropolitan area and assume that there are no regularly vacant dwellings and one household uses only one dwelling unit.

Figure 15.1 shows the change of households and dwellings in a regional housing market. Residential facilities that people live in are classified into dwellings and other occupied buildings. Dormitories, hotels, boarding houses and construction camps are included under other occupied buildings. Households are classified into principal households and lodging households. Hereafter, households indicate only principal households in this Chapter. Areas where households move to or from are classified into the Osaka prefecture and other areas.

Households who move to or from the regional housing market are classified as follows: (1) Z_{ij}: intra-region movers who move from housing type j to housing type i ($i = 1, \cdots, n$, $j = 1, \cdots, n$); (2) LH_j: departing households who move from housing type j within the region to an area outside the region, or lost households due to death or combining with another principal household; (3) ND_j: new demander for housing type i (households moving to the region from outside, newly formed households). Dwellings supplied in the market are classified as follows: (1) V_{ij}: available vacant dwellings left by intra-region movers Z_{ij}; (2) LS_j: available vacant dwellings left by departing households or lost households LH_j; (3) CS_j: newly constructed dwellings for housing type j.

Available vacant dwellings are part of the total vacant dwellings left by moving households. The rest of the vacant dwellings are considered to be demolished as we assume no regularly vacant dwellings. The ratio of the number of dwellings demolished to that of vacant dwellings left by movers is called the demolition rate and is represented by r_j for housing type j. Then the relation of movers and available vacant dwellings is written as follows (Kashiwadani, 1974):

$$V_{ij} = (1 - r_j)Z_{ij}, \tag{1}$$
$$LS_j = (1 - r_j)LH_j. \tag{2}$$

The number of dwellings demolished for type j, D_j, is shown by:

	Housing type	Intermediate demand 1	···	j	···	n	Final demand	Total demand
Available vacant dwellings left by intra-region movers	1	V_{11}	···	V_{1j}	···	V_{1n}	Y_1	X_1
	⋮	⋮		⋮		⋮	⋮	⋮
	i	V_{i1}	···	V_{ij}	···	V_{in}	Y_i	X_i
	⋮	⋮		⋮		⋮	⋮	⋮
	n	V_{n1}	···	V_{nj}	···	V_{nn}	Y_n	X_n
First supply		F_1	···	F_j	···	F_n		
Total supply		X_1	···	X_j	···	X_n		

Figure 15.2 : Housing input-output table

$$D_j = r_j(LS_j + \sum_i Z_{ij}). \tag{3}$$

Let us consider total demand and total supply over the regional housing market for a period. Total demand X_i for type i and the total supply X_j for type j are represented as follows respectively:

$$X_i = \sum_j Z_{ij} + ND_i = \sum_j V_{ij} + R_i + ND_i, \tag{4}$$

$$X_j = \sum_i V_{ij} + CS_j + LS_j \tag{5}$$

where $$R_i = \sum_j Z_{ij} r_j. \tag{6}$$

Following the terms of input-output analysis, we call V_{ij}, $Y_i(= R_i + ND_i)$ and $F_j(= LS_j + CS_j)$ respectively, the intermediate demand, the final demand and first supply. We can rewrite (4) and (5), with, Y_i, and F_j, as follows respectively:

$$X_i = \sum_j V_{ij} + Y_i, \tag{7}$$

$$X_j = \sum_i V_{ij} + F_j. \tag{8}$$

Using (7), and (8), we get the regional housing input-output table. This table is shown as Figure 15.2. Defining output coefficients p_{ij} by (9) and using matrix P. Equation (10) is derived from (7) and (8).

$$p_{ij} = \frac{V_{ij}}{X_i}, \sum_j p_{ij} + \frac{Y_i}{X_i} = 1, \tag{9}$$

$$X = P^* X + F \tag{10}$$

where

$$X = \begin{pmatrix} X_1 \\ \vdots \\ X_n \end{pmatrix}$$

$$F = \begin{pmatrix} F_1 \\ \vdots \\ F_n \end{pmatrix}$$

$$P = \begin{pmatrix} p_{11} & \cdots & p_{1n} \\ \vdots & & \vdots \\ p_{n1} & \cdots & p_{nn} \end{pmatrix}.$$

Further X is shown by

$$X = (I - P^*)^{-1} F. \tag{11}$$

As the number of available vacant dwellings of type j caused by F_j is $X_j - F_j$, D_j is shown by

$$D_j = r_j(\frac{X_j - F_j}{1 - r_j} + LH_j). \tag{12}$$

Using (11) and the definition of F_j, (12) is written as follows:

$$D_j = \frac{r_j}{1 - r_j}(((I - P^*)^{-1} - I)CS_j + (I - P^*)^{-1} LS_j). \tag{13}$$

The rate r_j to $(1 - r_j)$ means the ratio of the demolition rate to the available vacant dwellings rate. The first term in the bracket of the right hand side shows the number of available vacant houses left by movers, which is caused by new housing construction and the second term shows those caused by departing households or lost households. Therefore, when we know the value of matrix P and demolition rate r_j, we can estimate the number of dwellings demolished corresponding to the number of CS_j and LH_j.

Table 15.3 : Number of movers in and from Osaka prefecture and house constructional demolition

			intra regional movers; housing type(moved from)							
			owned		rented pub.	rented privately			issued houses	new demand
			D/T	A		D/T	A(W)	A(NW)		
housing type (moved into)	owned	D/T	86334	9716	36318	54241	14952	15027	14661	61226
		A	11296	8751	23536	14036	9749	9815	8401	28671
	rented pub		7074	1589	49110	34524	23658	20594	3021	45617
	R pri.	D/T	13445	2649	13207	45493	18833	15314	3430	59191
		A(W)	10402	2051	16088	20838	28013	22781	1891	93680
		A(NW)	10553	2075	13055	16707	25383	19026	1900	81834
	issued		1055	445	2785	4290	3608	3167	6962	46002
out movers			141011	27773	148847	190574	124298	105911	40273	
leaving or .lost			36265	7035	24500	22000	147044	15563	44726	
newly constructed			213519	97199	27548	9482	2728	80018	5558	
in movers			292859	114957	185969	172118	189650	175378	69172	
demolished			97268	16942	14123	48782	87626	25584	21856	

Note D/T: detached or tenement houses, A:apartment houses, A(W):wooden apartments, A(NW):non wooden apartments, R pub.: rented publicly, R pri.:rented privately

15.4 AN EMPIRICAL STUDY IN OSAKA PREFECTURE

The information in Table 15.3 is obtained using data from the Housing Survey of Japan and The Annual Statistical Report of Buildings. Here a data modification was performed by the author to weed out discrepancies between the number of households and the number of dwelling units reported by these two statistical publications. There are seven dwelling types classified by tenure and building type. The reasons that tenure is used as the main index to classify dwellings are follows: the first is that the number of movers is reported mainly in terms of tenure change in the Housing Survey of Japan, and the second is that tenure is a good approximate index of housing quality. Table 15.3 shows the number of movers between 1978 and 1983. The demolition rate r_j is calculated as the ratio of the number of dwellings demolished to the number of vacant dwellings left by movers, where the number of dwellings demolished in a period is calculated by sub-

Table 15.4 : Housing demolished rate

owned detached or tenement	0.5487
owned apartment	0.4867
rented publicly	0.0815
rented privately detached or tenement	0.2295
rented privately wooden apartment	0.3229
rented privately non wooden apartment	0.2106
issued houses	0.2571

tracting the amount of housing stock by type at the end of the period from the sum of the amount of housing stock at the start of the period and the number of dwellings built during the period, as we do not have good statistics concerning housing demolition.The number of vacant dwellings is the sum of those left by intra-region movers and those left by departing or lost households. Table 15.4 shows the demolition rate for each housing type. The rates of owned houses for both detached or tenement houses and apartments are larger than the rates of rented or issued houses. As for detached or tenement houses this is because the number of demolished dwellings includes houses demolished by owners who rebuild their new houses on the same sites. As for owned apartments it may be due to the discrepancy between two data sources, the Housing Survey and Annual Building Reports, because we can suppose cases where someone buys apartment houses and rents them to other people, and then the houses are reported as rented houses by the Housing Survey while they are reported as owned houses by the Annual Building Report. The demolition rates are more than 20% except for the rate of rented houses owned publicly.

A housing input-output table is obtained by subtracting the number of vacant dwellings demolished - which is calculated as a product of the number of movers and the demolition rate - from the number of movers shown in Table 15.3. The housing input-output table is shown in Table 15.5. It can be seen that construction and demolition patterns are very different among housing types. The number of dwellings

Table 15.5 : Housing input-output table

housing type (moved into)		available vacant dwellings left by intra regional movers (movers from)							demander leaving no available vacant houses
		owned		rented	rented privately			issued	
		D/T	A	pub.	D/T	A(W)	A(NW)	houses	
owned	D/T	38964	4987	33359	41794	10123	11862	10891	140879
owned	A	5098	4492	21618	10815	6601	7748	6241	52344
rented pub.		3193	816	45109	26601	16018	16257	2244	75731
R pri.	D/T	6068	1360	12131	35012	12751	12089	2548	90159
R pri.	A(W)	4695	1053	9817	16056	18967	17983	1405	119674
R pri.	A(NW)	4763	1065	11991	12873	17186	15019	1411	111070
issued		747	228	2558	3306	2443	2500	5172	52218
AVD(intra movers)		63641	14255	136720	146841	84158	83605	29918	
AVD(leaving, lost)		16367	3611	22504	16951	99558	12285	33225	
newly constructed		213519	97199	27548	9482	2728	80018	5558	
total supply		292859	114957	185969	172118	189650	175378	69172	
demolished		97268	16942	14123	48782	87626	25584	21856	

Note D/T: detached or tenement, A:apartments, A(W):wooden apartments,
A(NW):non wooden apartments, R pub.:rented publicly, R pri.:rented privately,
AVD(intra movers):available vacant dwellings left by intra region movers,
AVD(leaving, lost):available vacant dwellings left by leaving or lost households,

demolished for owned detached or tenement houses is fairly large but the number of dwellings constructed is much larger. The rate of constructed dwellings compared to demolished dwellings is large for non wooden apartments, both owned and privately rented. The number of dwellings constructed is much less than the number of dwellings demolished for the three housing types, privately rented detached or tenement houses, privately rented wooden apartments and issued houses. In particular, the rate of dwellings demolished compared to dwellings constructed for privately rented wooden apartments is large. It is more than 30 times larger. This means that this type of house is the most inferior of the housing types and these houses are being driven out of the market by dwellings of better quality construction.

Table 15.6 : Matrix P^*

Housing type	1	2	3	4	5	6	7
1	0.1330	0.0443	0.0172	0.0353	0.0248	0.0272	0.0108
2	0.0170	0.0391	0.0044	0.0079	0.0056	0.0061	0.0033
3	0.1139	0.1881	0.2426	0.0705	0.0518	0.0684	0.0370
4	0.1427	0.0941	0.1430	0.2034	0.0847	0.0734	0.0478
5	0.0346	0.0574	0.0861	0.0741	0.1000	0.0980	0.0353
6	0.0405	0.0674	0.0874	0.0702	0.0948	0.0856	0.0361
7	0.0372	0.0543	0.0121	0.0148	0.0074	0.0080	0.0748

Table 15.7 : Matrix $[I - P^*]^{-1}$

Housing type	1	2	3	4	5	6	7
1	1.1779	0.0786	0.0514	0.0668	0.0477	0.0501	0.0233
2	0.0257	1.0474	0.0120	0.0146	0.0105	0.0110	0.0061
3	0.2297	0.3194	1.3886	0.1610	0.1187	0.1392	0.0776
4	0.2803	0.2277	0.2973	1.3277	0.1686	0.1576	0.0971
5	0.1063	0.1380	0.1806	0.1443	1.1556	0.1536	0.0665
6	0.1109	0.1460	0.1786	0.1375	0.1477	1.1387	0.0661
7	0.0582	0.0748	0.0287	0.0292	0.0173	0.0181	1.0858

1 :owned detached or tenement
2 :owned apartment
3 :rented publicly
4 :rented privately détached or tenement
5 :rented privately wooden apartment
6 :rented privately non wooden apartment
7 :issued

Table 15.8 : The number of dwellings demolished caused by original housing supply

			originally supplied dwellings	number of dwellings demolished: owned D/T	owned A	rented pub.	rented privately D/T	rented privately A(W)	rented privately A(NW)	issued houses	total
owned	D/T	C	213519	46182	5203	4350	17829	10824	6318	4301	95007
		L	16367	23439	399	333	1367	830	484	330	27182
	A	C	97199	9289	4368	2754	6593	6397	3786	2516	35703
		L	3611	345	3586	102	245	237	141	93	4749
rented publicly		C	27548	1722	314	950	2440	2373	1313	274	9386
		L	22504	1407	256	2772	1993	1938	1072	224	9662
rented privately	D/T	C	9482	770	131	135	926	652	348	96	3058
		L	16951	1376	234	242	6705	1166	622	171	10516
	A(W)	C	2728	158	27	29	137	202	108	16	677
		L	99558	5774	991	1048	5000	54867	3923	596	72199
	A(NW)	C	80018	4874	834	988	3757	5862	2961	501	19777
		L	12285	748	128	152	577	900	3732	77	6314
issued houses		C	5558	158	32	38	161	176	98	165	828
		L	33225	941	192	229	961	1053	586	12486	16448
total		C	436052	63152	10910	9244	31842	26486	14931	7869	164434
		L	204501	34030	5787	4878	16847	60992	10560	13977	147071

Note D/T:detached or tenement, A:apartments, A(W):wooden apartments, A(NW):non wooden apartments, rent pub.:rented publicly, rent pri.: rented privately, C:dwellings supplied by new construction, L:available vacant dwellings left by new departing or lost households.

Matrix P^* and $(I-P^*)^{-1}$ are shown in Tables 15.6 and 15.7 respectively. Each $i-j$ component of matrix $(I-P^*)^{-1}$ shows the average number of housing vacancies in type i caused by one originally supplied dwelling unit of type j. Here the originally supplied dwellings mean dwellings newly constructed or dwellings left by departing or lost households. Table 15.8 shows the number of dwellings demolished in type j caused by the original housing supply in type i. It can be seen that owned houses - both detached or tenements - and apartments are the source of many demolished dwellings. The number of demolished dwellings due to owned detached or tenement house construction is particularly large. Two reasons are considered. The first is that the number of dwellings of the type constructed is large and the second is that dwellings of the type constructed draw out more housing vacancies. We can see the average number of available vacant dwellings caused by the original housing supply by house type as the sum of the elements of matrix $(I - P^*)^{-1}$ along a column in the matrix. The numbers are larger for owned houses or publicly rented houses than for privately rented houses or issued houses. Detached or tenement houses rented privately represent one of the housing types which have recently decreased in stock. In this case the number of dwellings demolished is mainly affected by newly constructed privately owned houses. That is, households who live in privately rented houses move to newly constructed privately owned houses and some of the empty rented houses are demolished. On the contrary, as for the number of dwellings in wooden apartments demolished, whether rented privately or of the issued houses type, the main reason for demolition is departing or lost households. Most residents in privately rented wooden apartments are young people, including students. Table 15.8 suggests that these apartment houses are demolished after they leave. Residents in issued houses are less connected with other housing types in the housing market. Therefore, the number of dwellings demolished in issued houses is not so affected by the housing supply of the other types.

If the numbers of matrix P^* and the demolition rate r_j are stable we can estimate the number of dwellings demolished by housing type according to the number of newly constructed dwellings and the number of departing or lost households which are both forecast beforehand. An estimation of the number of dwellings demolished between the period 1983-1988 is attempted using the same matrix P^* and demolition rate r_j as that from 1978 to 1983. It is also assumed that each number of departing or lost households by housing type is identical to that between 1978 and 1983 because the inter-regional migration pattern of the Osaka Prefecture is stable during these years. The number of newly constructed dwellings is obtained from recent Annual Statistical Construction Reports. The estimate of the number of houses demolished is calculated by Equation (13) using these data. The actual numbers of dwellings demolished is derived by subtracting the total housing stock in 1988 from the sum of the value of housing stock in 1983 and the number of dwellings newly constructed during the period. This is shown in Table 15.9. The estimated total number of demolished dwellings is a little larger than the actual one. The estimate of demolished dwellings is calculated under the assumption that the housing market situation is same as that between 1978 and 1983. However, between 1983 and 1988 the market was tighter than that from 1978 to 1983. The number of households increased by 195,000 during

Table 15.9 : Number of dwellings demolished actual and estimated (1983-1988)

Housing type	actual	estimated
1	105159	87368
2	8127	15302
3	5423	13452
4	33505	48688
5	63977	90764
6	33830	26765
7	27667	20627
Total	277688	302906

1 :owned detached or tenement
2 :owned apartment
3 :rented publicly
4 :rented privately detached or tenement
5 :rented privately wooden apartment
6 :rented privately non wooden apartment
7 :issued houses

the period 1983-1988 compared with 122,000 between 1978 and 1988. On the other hand there were 458,000 newly constructed houses from 1983 to 1988, which is almost same as the number built during the period 1978-1983: 436,000. Therefore, the relatively higher demand than supply during the period 1983-1988 saved more vacant dwellings and reduced the number of demolished dwellings. The over-estimate of demolition for privately rented wooden apartments - which is the lowest quality housing type - coincides closely with this observation.

We project the number of houses in 1988 by housing type using the value of housing stock in 1983, the actual number of houses constructed between 1983 and 1988, and the estimated number of houses demolished using the model. Table 15.10 shows the actual housing stock and the projected stock in 1988 by housing type. Each estimated number looks close to its corresponding actual number. The average value of percentage error over housing types is 4.4%. It is much less than that of Table 15.9, 55%. We can get fairly good projection results for the housing stock while the estimated number of houses demolished is not so close to the actual numbers. This is because the volume of housing stock reserved is much larger than that demolished during the period. This is a merit of incremental models compared to stock allocation models.

Table 15.10 : Total housing stock, 1988 actual and projected

Housing type	actual	projected
1	1180551	1198342
2	233957	226782
3	353825	345796
4	266428	251247
5	299102	272315
6	379034	386099
7	92102	99142
Total	2804999	2779723

1 :owned detached or tenement
2 :owned apartment
3 :rented publicly
4 :rented privately detached or tenement
5 :rented privately wooden apartment
6 :rented privately non wooden apartment
7 :issued houses

15.5 CONCLUDING REMARKS

When we are faced with the long term projection of the urban housing market or a location problem we immediately notice that dwellings of some types are increasing in number and other types are decreasing. As housing is a typical stock good and its average life span is fairly long, the future dynamic change of housing depends not only on future housing demand and supply but also on the present housing stock distribution by age or housing type. Therefore, we can get useful information from the present housing stock in a city to project the future situation. However, it is very difficult to predict how the present housing stock will be changed by the future demand and supply. On the other hand, the present housing stock may affect future construction levels. The interaction between housing stock and housing flow - both of demand and supply - is very complicated. This Chapter provides a first step in achieving an understanding of this complicated process.

Only one device for framing the various moves of households and houses in a urban area is shown. There is much work to do in the future. In particular we need a theoretical model to analyze the interaction between housing stock and housing flow.

REFERENCES

Ingram, G.K., Kain, J.F. and J.R. Ginn, 1972, *The Detroit Prototype of the NBER Urban Simulation Model*, National Bureau of Economic Research, Chicago.

Kashiwadani, M. ,1974, "A study on housing demand-supply models in metropolitan areas", *Proceedings of Japan Society of Civil Engineers*, 227:61-69 (in Japanese).

Kashiwadani, M. and A. Kawachi, 1989, "A study of factors of housing demolition", *Proceedings of Infrastructure Planning*, 12:635-641 (in Japanese).

Putman, S.H., 1975, "Urban land use and transportation models, A state-of-the art summary", *Transportation Research*, 9:187-202.

Rothenberg, J., Galster, G.C., Butler, R.V. and J. Pitkin, 1991, *The Maze of Urban Housing market*, The University of Chicago Press, Chicago and London.

Wegener, M., 1982, "Modeling urban decline, A multi level economic-demographic model for the Dortmund region", *International Regional Science Review*, 7:2:217-241.

Wissen, L.V. and A. Rima, 1988, *Modelling Urban Housing Market Dynamics*, Elsevier Science Publishers B.V., New York.

CHAPTER 16

Optimal Regional Investment Control Using Hallmark Events

Makoto Okumura, Kazuhiro Yoshikawa
and Eizo Hideshima

16.1 INTRODUCTION

During the 80's, the main objective of regional development policy in Japan gradually shifted from the creation of facilities for industrial activities to the creation of a comfortable atmosphere for leisure and recreation activities. The Comprehensive Recreation Area Development Act of 1987 and the Special Fund for Local Community Development of 1988/1989 eloquently reflect this objective shift and the growing importance of software in regional development. The great variety of hallmark events that has come to be introduced into regional development projects is related to this shift.

Ritchie (1984), one of the major contributors to the analysis of hallmark events, defined hallmark events as the following: 'Major one-time or recurring events of limited duration, developed primarily to enhance the awareness, appeal and profitability of a tourism destination in the short and/or long term. Such events rely for their success on uniqueness, status, or timely significance to create interest and attract attention'. He understood hallmark events to be 'major' events which have an ability to focus national and international attention on the destination; his definition tends to be dependent upon the 'scale' of the event.

However, other researchers such as Shepherd (1982) have pointed to the significance of small scale entertainment events - such as carnivals, festivals and fetes, in small towns and villages - which, in relation to their regional setting, may well be described as hallmark tourist events. Such an observation notes the importance of the context within which hallmark events take place. A hallmark event such as an EXPO, exhibition, trade fair, convention, symposium, festival, cultural event, sporting competition, e.t.c., is the most effective means for announcing the superiority or characteristics of the host region (Hall, 1989a). Advertising messages are spread widely via several types of media. In many cases the event not only has a significant impact on the inhabitants of the host community but also on the visitors and tourists. The people who take part in such an event will also convey their impressions of the host region to other people by oral media; its reputation will increase with the cumulation of visitors. A good reputation has great power to increase the number of visitors to an area, which induces an expansion of the monetary inflow to the regional economy and a quicker return on capital. Moreover, redistributions of income, employment and investments are also important effects of hallmark events. Considering the merits of such events, the strong motivation to introduce them into regional projects, especially resort development, can easily be understood.

Currently, the most dominant methodology used in regional planning is probably 'demand oriented planning', which is structured in order to determine the best way to satisfy existing, or constantly increasing demands. However, the introduction of hallmark events into the arena of regional development requires the development of a new planning methodology. Visitors to hallmark events decide the reputation of the region. According to this reputation many other people will decide whether they will visit the region or not. Thus, this is an autonomous demand creation process. In this situation a priori forecasting of demand is almost impossible. We must develop an on-going control method. This study may contribute towards this challenge to some extent.

16.2 RELATED STUDIES TO DATE

A hallmark event is one of the most effective ways to promote the characteristics of a region, especially in relation to a resort development project. Hallmark events and resort development policies have been studied in the field of tourism management. There are many experiential reports; some are admiring reports of successful projects, while others are critical reports of failed projects, but they all lack a well established framework. Ritchie (1984) proposed a classification of the types of impacts which need to be assessed for a hallmark event: economic, tourism/commercial, physical, socio-cultural, psychological and political. He also noted that his classification was only a beginning and that there was a need for a 'more comprehensive' approach to the assessment of the impact of hallmark events than that currently employed in most situations. That evaluation is still relevant today.

Economic impact analysis is the most advanced part of hallmark event impact analysis, because attention has been concentrated on obvious economic impacts in the field of tourism research as mentioned by Hall (1989b). Most of the reported studies on the impacts of a hallmark event have tried to measure the economic impacts of a specific event. According to Roberts & McLeod (1989), these studies are usually based on one of two approaches. The first is an economic impact study which examines the effect of the event on output, household income and employment in the host region. Input-output analysis is popular in such studies. The second approach is social cost benefit analysis which examines the net economic benefits of the event to the host community. While an economic impact study is a planning tool which is carried out prior to the event, social cost benefit analysis is usually carried out after the event when a full accounting of the benefits and costs is possible.

Whether studies are based on the first or second approach, they almost always neglect the fact that effects of hallmark events are closely related to infrastructure. Schaer (1978) pointed out the importance of the synchronization of the planning of hallmark events with infrastructure development. 'The hosting of a hallmark event without the provision of an adequate infrastructure to cope with increased visitor numbers may well cause a decline in urban environmental quality, for instance, in the impacts of increased traffic flows'. Investment budgets and the stock of infrastructure - therefore, the marginal effect of a certain size event - may change along the axis of time. The close relationship between hallmark events and infrastructure makes it difficult to plan such events independently.

In order to analyze the effects of hallmark events with linkages to infrastructure development, a dynamic and normative approach is required. The calculus of variations and Pontryagin's optimal control theory provide strong tools for dynamic normative analysis. Some important applications of optimal control theory are employed in regional economics; see e.g. Miyao (1987), Bennett & Hordijik (1986). About a quarter of a century ago, optimal regional growth paths were analyzed with optimal control theory by Koopmans (1965), Gale (1967) and Arrow (1968). Control theory was also applied to the multi-sector Harrod-Domer growth problem. Rahman (1963, 1966) suggested an investment allocation model which contains two regions interacting with each other. These regional optimal investment models were reviewed and classified by Fujita (1978); however, they were developed as theoretical models which provide no means of empirical calibration.

A resolution was provided by Tan and Bennett (Tan, 1979; Tan &

Bennett, 1984); control theory was applied to empirically calibrated econometric systems. Their methodology provided a way in which to combine empirical and normative models. However, it was only applicable to linear equation systems. That restriction was problematic for regional economic systems which contain non-linear mechanisms. In this study we will deal with an optimal control problem of the nonlinear system expressed by a nonlinear logit model. The logit model was developed by McFadden as a standard type of qualitative choice model based on random utility, and is now widely used for modelling inter-regional linkages. The findings of this study may nourish analyses of regional control problems.

16.3 INVESTMENTS IN HALLMARK EVENTS AND INFRASTRUCTURE

In this study we divide the numerous types of regional investments into two categories; investments in hallmark events and in infrastructure. In reality, large scale hallmark projects, such as international EXPOs, almost always contain numerous infrastructure development projects. However, for the ease of theoretical analysis, we will distinguish between them according to the duration of project's effect. We define investments in hallmark events as investments whose impact is only apparent during the duration of the event. Contrarily, investments in infrastructure are such that their utility lasts a long time after the investments are made.

At present, most regional investment by the public sector is categorized as infrastructure investment. However, the percentage of hallmark events will certainly increase, reflecting the trend from hardware to software. Hallmark events are decisive in immediately enlarging the number of visitors or tourists to the host region, but this phenomenon will not be cumulated, because it damps out instantly after investment ceases. Therefore, hallmark events are usually designed with a short-term scope.

Infrastructure, the other category of investment, is planned in a long-term perspective: because facilities cannot be easily scrapped and rebuilt, once they have been constructed they have to be used for a long time. The effect of investments is cumulated and forms the basic level of regional comfort or attractiveness. Considering the slow return on capital, investments in infrastructure must endure for a long time.

It is most important to recognize that the effect of hallmark events is not only dependent on the quality and scale of the event, but also on its accompanying regional infrastructure stock. Let us consider an event such as a traditional festival in a region. Without any transportation or organizational support, it is difficult for people to travel

Table 16.1 : Hallmark events v.s. Infrastructure

Investment for	Hallmark Events:E(t)	Infrastructure:F(t)
Definition	Investment whose effect is obtained only while it is executed.	Investment in facilities which last a long time
Current amount in projects	Small (Supplemental)	Large (Main)
Power to enlarge the number of visitors	Large Instant Temporary	Small Steady Cumulative
Time-span	Short-run	Long-run
Merits	Quick Return Means to display Regional Identities	Cumulative Basis for Regional Attractiveness
Drawbacks	Negative Reputation generated if too crowded	Slow Return

there and to enjoy that event. Thus, the number of visitors or tourists is not only dependent on uniqueness, ideas, know-how and the timing of the hallmark event, but also on the fundamental infrastructure stock for the visitors' use.

Our main problem is how to arrange these two types of investments under budget constraints. If they are arranged efficiently the region may perform more highly, there may be more visitors and increased economic effects. In many big projects, a hallmark event is held to coincide with the completion of an infrastructure development. For example, a hallmark event entitled Portpia '80 was held when the Port Island reclaimation project in Kobe was completed. It would be interesting to study when such events should be held.

The main cash flows and relationships of regional development projects are illustrated from an economic point of view in Table 16.1. Investments are executed for hallmark events and regional infrastructures under limited budget constraints. The number of visitors and tourists is dependent upon the level of investment in the event and the volume of infrastructure stock, as well as the reputation of the region. The consumption of visitors and tourists decides the inflow of resources into the regional economy. In order to best utilize hallmark

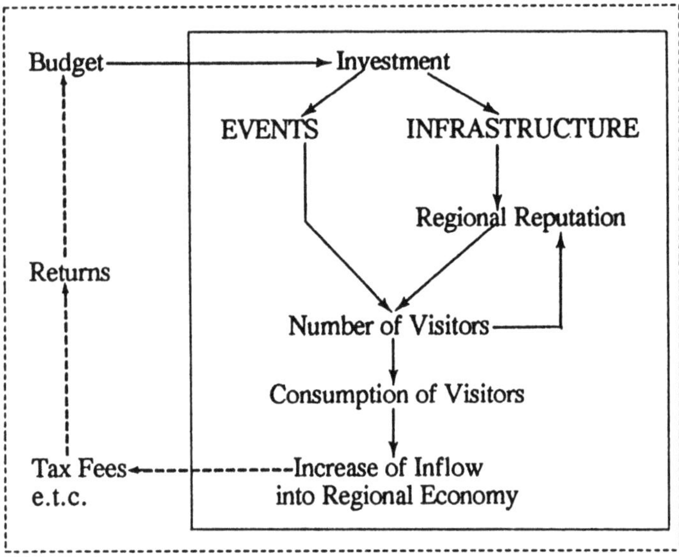

Figure 16.1 : Economic flows in a regional development project

events and the resulting increase of tourist consumption as additional economic bases for regional development, we must try to increase the number of visitors attracted by the hallmark events.

Here we can set the first prototype of the control problem. Though hallmark events are important for investment planning, the budget must be shared with infrastructure investment. While many visitors are attracted to a region by hallmark events, a sufficient infrastructure capacity is indispensable; otherwise, unsatisfied visitors will spread bad reports about the region, resulting in a subsequent decrease of visitors. Infrastructure must be improved in order to satisfy the needs of the increased number of visitors caused by the events. In other words, hallmark events can only be planned with accompanying improvements in infrastructure. In this Chapter, this problem will be attacked with the aid of a strong mathematical tool: optimal control theory.

In Figure 16.1 there is a feedback loop; inflows into the regional economy will alter the scale of the investment budget in the future. This feedback linkage provides the potential for quick returns on those regional investments via resources actually realized by some type of taxation, entry fees etc. Considering this total system with its feedback loop, it is obvious that the total amount of investment can be shifted according to the number of visitors attracted to the event. Here, the second type of investment control problem can be set. The question is how to expand and improve infrastructure using the quick capital return provided by hallmark events. This is also an interesting issue for some future study.

16.4 BEHAVIORAL MODEL OF TOURISTS' DESTINATION CHOICE

It is assumed that several destinations are competing for a fixed number of tourists. Each tourist evaluates the attractiveness of alternative destinations and chooses the most attractive alternative before departure. The attractiveness or utility of each alternative may differ between individuals because of their variety of tastes. According to random utility theory, the attractiveness or utility of alternative j, indicated by U_j, is described as the sum of the systematic parts, $\bar{U}_j$, which is constant over all individuals, and the random part, ε_j. Here, the systematic parts are assumed to be an additive form of event attractiveness V_j and infrastructure attractiveness W_j,

$$U_j = \bar{U}_j + \varepsilon_j = V_j + W_j + \varepsilon_j. \tag{1}$$

When the random part, ε_j, obeys the mutually independent Gumbel distribution, the number of tourists of alternative j, indicated by P_j, is estimated according to the following multinominal logit model,

$$P_j = Q\frac{\exp \bar{U}_j}{\sum_k \exp \bar{U}_k}, \tag{2}$$

where Q is the total number of tourists whose choice set contains the focused region j.

Here, event attractiveness is assumed to be a function of the intensity of the event investment $E_j(t)$, measured in monetary terms. Undoubtedly, uniqueness or excellent ideas have the power to enlarge the effect of the same amount of event expenditure. That is beyond our scope here; however, such a discussion would be indispensable to calibrate the value of parameter α in a more empirical study.

$$V_j = \alpha E_j(t). \tag{3}$$

Let us discuss the attractiveness of infrastructure. In general, information about the stock of infrastructure is conveyed less often by the mass-media than information about a hallmark event. The utility of infrastructure is also dependent upon the level of congestion. Consequently, as tourists cannot observe the actual status of the infrastructure stock before they visit a region, they sometimes choose their travel destination on the basis of the 'reputation' of its infrastructure. We describe this 'reputation' $W_j(t)$ as a cumulation of the 'respect' of the visitors who have already visited that place, $R_j(t)$,

$$W_j(t) = W_j(0) + \int_0^t R_j(\tau)d\tau. \tag{4}$$

The term respect describes the satisfaction level of a visitor who has actually visited and used the infrastructure of the region, explained by a positive term for the infrastructure stock and a negative term for over-crowding, as follows:

$$R_j(t) = \beta S_j(t) - \gamma P_j(t). \tag{5}$$

The stock of infrastructure, $S_j(t)$, represents the total amount of infrastructure investment $F_j(t)$,

$$S_j(t) = S_j(0) + \int_0^t F_j(\tau)d\tau. \tag{6}$$

In the next Section, our analysis will focus on the control of a certain destination j; therefore, subscript j will be omitted without notice.

16.5 INVESTMENT CONTROL MAXIMIZING THE NUMBER OF VISITORS

In this Section, we analyze the first prototype of the control problem: how do we utilize hallmark events in order to maximize the number of visitors under fiscal constraints? We consider a situation in which only a focused region is controllable; the infrastructure level and utility of the other competing destinations are assumed to be constant. This assumption may be a little strong in reality, and it should be relaxed in future studies. Under this assumption, the utility level of the destinations other than the focused region can be synthesized in the form of the log-sum variable, L; the number of visitors of the focused region can be calculated by the following logit model,

$$P(t) = Q\frac{\exp \bar{U}(t)}{\exp L + \exp \bar{U}(t)} = Q\frac{\exp W(t) \exp V(t)}{\exp L + \exp W(t) \exp V(t)}. \tag{7}$$

The total number of visitors in the given planning period, e.g., five or ten years, can be set as an objective function of this control problem:

$$\text{Maximize } J = \int_0^T P(\tau)d\tau. \tag{8}$$

The budget for investment is assumed to be constant during the planning period; C indicates the amount of the budget available for one unit of time,

$$F(t) + E(t) = C. \tag{9}$$

In this situation, the proportion of event investment to the total budget, $u(t)$ $(0 \le u(t) \le 1)$, can be defined as a control variable.

$$E(t) = Cu(t) \tag{10}$$

$$F(t) = C(1 - u(t)). \tag{11}$$

By optimal control theory, as explained in Leitmann (1966) and Takayama (1985), etc., this optimal control problem is converted to the Hamiltonian maximizing problem of the following system:

$$x_0 = J(t), \quad \frac{dx_0}{dt} = Q\frac{\exp W(t) \exp V(t)}{\exp L + \exp W(t) \exp V(t)} \tag{12}$$

$$x_1 = W(t), \quad \frac{dx_1}{dt} = \beta S(t) - \gamma p(t) \tag{13}$$

$$x_2 = S(t), \quad \frac{dx_2}{dt} = C(1 - u(t)) \tag{14}$$

$$x_3 = t, \quad \frac{dx_3}{dt} = 1. \tag{15}$$

Costate functions, $\psi_0, \psi_1, \psi_2, \psi_3$, are introduced to formulate the Hamiltonian H, to be maximized:

$$H = -\psi_0 P(t) + \psi_1(\beta S(t) - \gamma P(t)) + \psi_2 C(1 - u(t)) + \psi_3. \tag{16}$$

Costate functions obey the following differential systems:

$$\frac{d\psi_0}{dt} = -\frac{\partial H}{\partial x_0} = 0 \tag{17}$$

$$\frac{d\psi_1}{dt} = -\frac{\partial H}{\partial x_1} = (\psi_0 + \gamma\psi_1)P(t) \tag{18}$$

$$\frac{d\psi_2}{dt} = -\frac{\partial H}{\partial x_2} = -\beta\psi_1 \tag{19}$$

$$\frac{d\psi_3}{dt} = -\frac{\partial H}{\partial x_3} = 0. \tag{20}$$

These differential equations are solved as follows:

$$\psi_0(t) = c_0 = -1 \tag{21}$$

$$\psi_1(t) = \frac{1}{\gamma}\{c_1 \exp(\gamma \int_0^t P^*(\tau)d\tau) - c_0\}. \tag{22}$$

A transversality condition, $\psi_1(T) = 0$, determines c_1.

$$c_1 = -\exp(-\gamma \int_0^T P^*(\tau)d\tau) < 0 \tag{23}$$

$$\frac{d\psi_1}{dt} = c_1 \exp(\gamma \int_0^t P^*(\tau)d\tau)P(t) \leq 0. \tag{24}$$

Therefore, $\psi_1(t)$ must be a monotonous decreasing function as time goes by. This means:

$$0 = \psi_1(T) \leq \psi_1(t) \leq \psi_1(0) = \frac{c_1 + 1}{\gamma} < \frac{1}{\gamma}. \tag{25}$$

On the other hand,

$$\psi_2(t) = c_2 - \int_0^t \beta\psi_1(\tau)d\tau \tag{26}$$

where, $\psi_2(T) = 0$ assures:

$$c_2 = \beta \int_0^T \psi_1(\tau)d\tau < 0 \tag{27}$$

$$\psi_3(t) = c_3 = 1. \tag{28}$$

If we substitute these results into Equation (15), the Hamiltonian can be derived as:

$$H^* = (1 - \gamma\psi_1^*(t))P^*(t) + \beta\psi_1^*(t)S^*(t) + \psi_2^*(t)C(1 - u(t)). \tag{29}$$

Let us consider this Hamiltonian as a function of control variable $u(t)$, and check the optimal conditions; first and second derivatives are derived as following:

$$\frac{dH^*}{du} = \alpha CQ \frac{(1 - \gamma\psi_1^*)\exp(W^*(t) + V^*(t))\exp L}{\{\exp(W^*(t) + V^*(t)) + \exp L\}^*} - \psi_2^*(t) \tag{30}$$

$$\frac{d^2H^*}{du^2} = \alpha^* C^* \frac{(1 - \gamma\psi_1^*)\exp LP^*(t)}{\{\exp(W^*(t) + V^*(t)) + \exp L\}^*}\{\exp L - \exp(W^*(t) + V^*(t))\}. \tag{31}$$

As $(1 - \gamma\psi_1^*(t))$ is positive and followed by inequality (25), the sign of the R.H.S. of equation (31) is coincident to the sign of $\{\exp L - \exp(W^*(t) + V^*(t))\}$, the relative market share. When the focused destination has a market share of greater than fifty percent, this term becomes negative and then interior optimum solutions are obtained. Otherwise, the two boundary cases, $u = 1$ and $u = 0$ are the only candidates for the optimal solution, since function $H^*(u)$ is convex. Here, we define $G = H(u=1) - H(u=0)$ in order to compare the two boundary conditions. This comparison function G can be rewritten by gathering W to the L.H.S.:

$$f(W(t)) \equiv \frac{\exp W \exp V^*}{\exp W \exp V^* + \exp L} - \frac{\exp W}{\exp W + \exp L}$$
$$\begin{matrix} > \\ < \end{matrix} \frac{C\psi_2^*(t)}{Q(1-\gamma\psi_1^*(t))} \equiv K(t) \sim \begin{cases} u^*(t) = 1 \\ u^*(t) = 0. \end{cases} \tag{32}$$

Here, we introduce a variable ω, which is a positive, monotonous decreasing function of $W(t)$;

$$\omega \equiv \frac{\exp L}{\exp W(t)}. \tag{33}$$

Then $f(W(t))$ can be rewritten as $g(\omega)$, a function of ω:

$$f(W(t)) \equiv g(\omega) \equiv \frac{\exp V^*}{\exp V^* + \omega} - \frac{1}{1+\omega}. \tag{34}$$

The first derivative is given as follows:

$$\frac{dg}{d\omega} = \frac{(1-\exp V^*)(\omega^2 - \exp V^*)}{(\exp V^* + \omega)^2(1+\omega)^2}. \tag{35}$$

Because $(1 - \exp V^*) \geq 1$, based on the fact that V^* is positive, the sign of the L.H.S. of Equation (35) is opposite to the sign of $(\omega^2 - \exp V^*)$. Then function $g(\omega)$ increases for small ω, but decreases in larger ω. Based on the monotonous nature of ω, the function $f(W(t))$ is concluded to have the same shape as $g(\omega)$. It usually intersects the criteria function, $K(t)$ at the two points W_1^* and W_2^*. Optimal control will be given as follows:

$$\begin{cases} W(t) \leq W_1^* \\ W_1^* \leq W(t) \leq W_2^* \\ W_2^* \leq W(t) \end{cases} \sim \begin{cases} u^*(t) = 0 \\ u^*(t) = 1 \\ u^*(t) = 0. \end{cases} \tag{36}$$

This result reveals two important points: one is that effective hallmark events cannot be held with insufficient infrastructure; the other is that hallmark events are effective when there is room to expand their market share. A hallmark event is seen as an effective tool for a medium competitor, but this is not so for minor entrants or a monopolist. In other words, the reputation of infrastructure, especially its relative

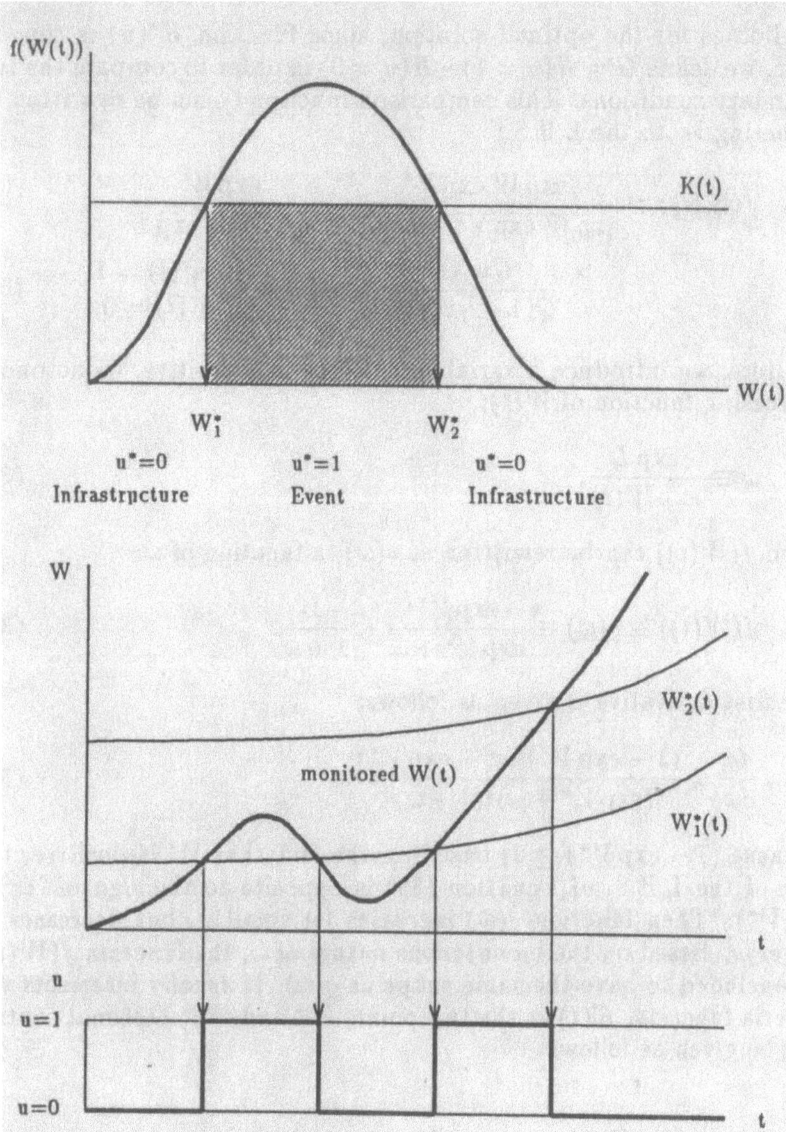

Figure 16.2 : Optimal switching procedure based on the criteria function $K(t)$

status compared to that of other competitors, has an important role in policy making.

Optimal control will be performed as follows. At first we solve the costate functions under the transversality conditions at the terminal period; using the assumed initial value, calculate these functions over time; after that, the initial value is adjusted until the transversality

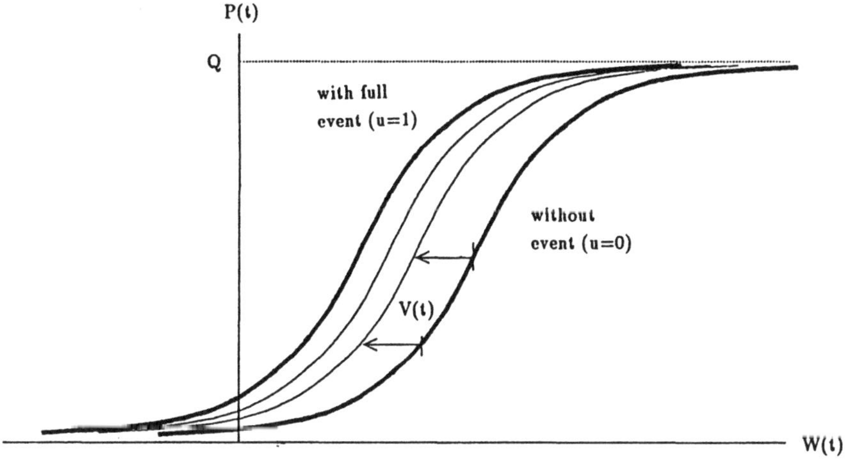

Figure 16.3 : Relationships between number of visitors and regional infrastructure reputation

conditions are met. Then we derive the criteria function $K(t)$, which indicates the switching point of 'bang-bang' control, W_1^* and W_2^*. The monitored value of $W(t)$ must be compared with those critical values and then switching is decided. In this procedure, monitoring the reputation of infrastructure is vital. This is illustrated in Figure 16.2.

16.6 DEVELOPMENT PATH OF REPUTATION

In this Section the dynamic pattern of development is assessed. The number of visitors increases as the reputation of a region increases. Figure 16.3 shows the S-shaped curves on the $W(t) - P(t)$ plane, formulated as a logit model in Equation (7). The right boundary curve indicates the relationship when no hallmark event is held. When a hallmark event is introduced, the curve shifts to the left along the length of $V(t)$; therefore, the left boundary curve is valid when all of the budget is utilized for the events. A family of S-shaped curves for various amounts of event investments makes a banana-shaped region, surrounded by the two boundary cases: $u(t) = 0$ and $u(t) = 1$.

Optimal control is designed in this feasible region on the plane. In order to attain our original objective, the maximization of the total number of visitors, we must design our control for both short and long-run perspectives. The short-run objective is the maximization of the number of visitors at any time, which is illustrated as a shift upwards in the figure. On the other hand, the long-run objective is the growth of reputation, illustrated as a shift towards the right in the figure. Now, let us investigate the process of reputation growth that has already been formulated in Equation (13), such that:

$$\frac{dW(t)}{dt} = \beta S(t) - \gamma P(t). \tag{37}$$

From this equation, the number of tourists which just balances the temporal stock of regional infrastructure can be calculated as follows:

$$P = \frac{\beta}{\gamma} S(t). \tag{38}$$

When the number of actual visitors $P(t)$ is smaller than this level P, reputation $W(t)$ increases. Contrarily, when $P(t)$ exceeds P, $W(t)$ decreases. A horizontal line of $P(t) = P$ can be drawn in the figure based on the current infrastructure stock. If event policy, $u(t)$, is kept constant, the status quo moves along the curve corresponding to the value of $u(t)$ and approaches the point where the horizontal line is crossed, while below the horizontal line, movement up towards the right can be realized along the curve. Once the status quo reaches the horizontal line, objectives can no longer be improved along the curve.

Here, we assume that the initial state is at point A in Figure 16.4. If we hold events, the status quo moves along the left boundary curve and stops at point B which is the intersection with the horizontal line. At that point, let us assume that the budget is concentrated for infrastructure development. The number of visitors decreases to point C, and after that the status quo moves along the right boundary. This time, however, the status quo does not stop at point D, the intersection with the same horizontal line, but at the upper point E. Because of the gradual investment, the level of infrastructure has shifted upwards, and saturation is postponed away from point D.

It is not possible to continue towards the right from point E, even if more visitors can be attracted by the hallmark event and the short-run objective is pursued. From point F, the status quo will sink along the left boundary curve and stop at point G. This converged situation is the same as that which occurred at point B.

According to these considerations, infrastructure investments (C to D) and event investments (F to G) may be executed one after the other. Strictly speaking, the switching occurs a little before the status quo has converged to the horizontal line; because the state changes very slowly near the horizontal line, a priori switching may give a better performance.

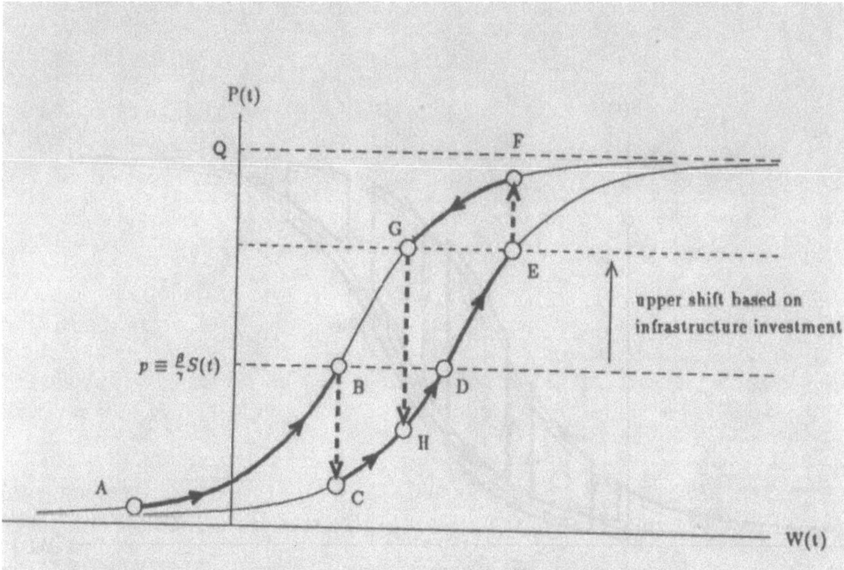

Figure 16.4 : Number of visitors corrresponding to current infrastructure stock and switchings

The important point is that the cycle is gradually shifted to the right due to the cumulated effect of infrastructure investment, and that the development path is a spiral, which is illustrated in Figure 16.5. In this spiral, infrastructure investment contributes to moves towards the right, which implies long-run scope reputation development; the hallmark event causes upward movement, or a temporal expansion of the number of visitors.

If we consider the competitive situation, the discussion becomes more complicated. Some research articles reveal bifurcating properties in a competitive logit system. It would be an interesting but tough task to analyze such control problems on the manifold with bifurcation.

16.7 CONCLUSION

Hallmark events, which yield quick harvests in the tourism market, have widely come to be introduced into planning for regional development as well as infrastructure investment. This Chapter aimed to model the regional investment process related to hallmark events and to obtain the optimal investment policy with the aid of optimal control theory.

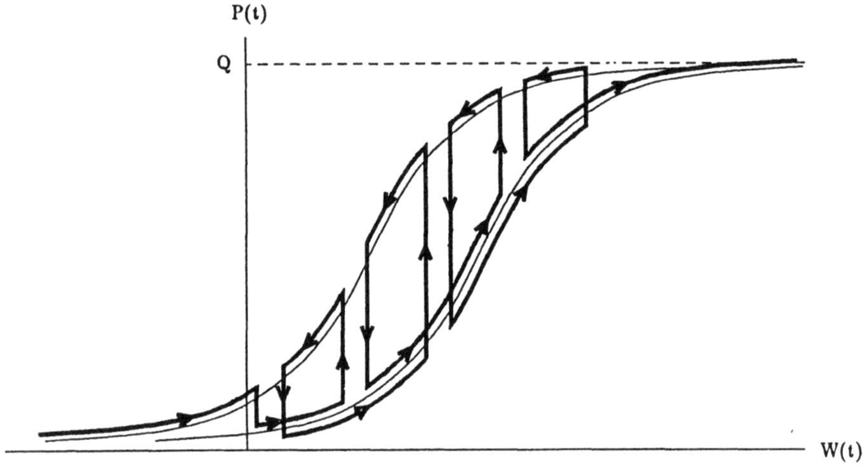

Figure 16.5 : Spiral development pattern

In the introduction, the definition of hallmark events was discussed and the socio-economic background for the strong motivation to introduce hallmark events was assessed; the need for a new methodology to replace the demand oriented one was emphasized. Existing research articles about the impact of hallmark events were reviewed in the second Section. To date they have often neglected the close relationship between the impacts of the events and infrastructure. Studies of regional development policy based on optimal control theory were shown to be very helpful for a dynamic normative approach. In Section 16.3, the synchronization of the planning of hallmark events with infrastructure development was discussed, mainly from an economic point of view. The behaviour of tourists in choosing their destination was formulated in Section 16.4; a logit type model which is popular in many transport studies was used for this purpose. In this study, however, the cumulated respect of visitors for a region is assumed to be very influential in determining the number of visitors thereafter, and is formulated as a reputation: that is a unique feature of this study. The following Section was based upon optimal control theory. The optimal investment policy was revealed to differ due to states of infrastructure reputation. The efficiency of the investments in hallmark events is closely dependent upon the relative superiority of infrastructure, or market-share. It was shown that a monitoring activity is indispensable in the optimal control procedure. This finding may imply that monitoring and on-going management must be introduced in the post-demand-oriented society in which a priori forecasting is not available. In Section 16.6, the optimal investment pattern was assessed graphically; a compromise between long-run and short-run objectives yielded a spiral development path.

It is hoped that this study and its findings may provide many useful suggestions for planners. However, it only represents the first stage of our investigations; there are many additional interesting issues for further studies. First, the competitive situation must be analysed more carefully with the aid of dynamic game theory. This may be a tough task involving the nonlinear analysis of bifurcated manifolds. Second, uncertainty in hallmark events must be considered in the analysis. The uniqueness of the events represents one side of the coin, but a high degree of uncertainty may represent the other. Such a discussion is also dependent on the quality of the hallmark event. Empirical studies which assess the validity and applicability of this study should also be undertaken. A new methodology may be needed in order to measure the realistic value of the parameters in the model suggested here.

REFERENCES

Arrow, K.J., 1968, "Applications of Control Theory to Economic Growth", in G.B. Dantig, et al. (eds.), *Mathematics of the Decision Sciences, Pt.2*, American Mathematical Society, pp. 85-119.

Bennett, R.J. and L.Hordijk, 1986, "Regional Econometric and Dynamic Models", in P. Nijkamp (ed.), *Handbook of Regional and Urban Economics, Vol. 1*, North-Holland, Amsterdam, pp. 407-441.

Fujita, M., 1978, *Spatial Development Planning*, North-Holland, Amsterdam.

Gale, D., 1967, "On optimal development in a multi-sector economy", *Review of Economic Studies*, 34:1-8.

Hall, C. M., 1989, "Hallmark Tourist Events: Analysis, Definition, Methodology and Review", in G.J. Syme, et al. (eds.), *The Planning and Evaluation of Hallmark Events*, Avebury, Aldershot, pp. 3-19.

Hall, C.M., 1989a, "Hallmark Events and the Planning Process", in G.J. Syme, et al. (eds.), *The Planning and Evaluation of Hallmark Events*, Avebury, Aldershot, pp. 20-39.

Koopmans, T.C., 1965, "On the Concept of Optimal Economic Growth", in *The Econometric Approach to Development Planning, Pontif-*

icia Academia Scientiarvm Scrlptvm Varia, North-Holland, Amsterdam, pp. 225-289.
Leitmann, G., 1966, *An Introduction to Optimal Control*, McGraw-Hill, New York.
Miyao, T., 1987, "Dynamic Urban Models", in E.S. Mills (ed.), *Handbook of Regional and Urban Economics, Vol. 2*, North-Holland, Amsterdam, pp. 877-925.
Rahman, M.A., 1963, "Regional allocation of investment", *Quarterly Journal of Economics*, 72:26-39.
Rahman, M.A., 1966, "Regional allocation of investment: Continuous version", *Quarterly Journal of Economics*, 75:159-160.
Roberts, E.J. and P.B. McLeod, 1989, "The Economics of a Hallmark Event", in G.J. Syme et al. (eds.), *The Planning and Evaluation of Hallmark Events*, Avebury, Aldershot, pp. 242-249.
Ritchie, J.R.B., 1984, "Assessing the impact of hallmark events: Conceptual and research issues", *Journal of Travel Research*, 23:1:2-11.
Schaer, U., 1978, "Traffic problems in holiday resorts", *Tourist Review*, 33:9-15.
Shepherd, D., 1982, "Community Organizations and Hallmark Events, a Case Study of the Wellesley Apple Butter and Cheese Festival", Research Report of Department of Recreation, University of Waterloo.
Takayama, A., 1985, *Mathematical Economics, 2nd Edition*, Cambridge University Press, Cambridge.
Tan, K.C., 1979, "Optimal control of linear econometric systems with linear equality constraints on the control variables", *International Economic Review*, 20:253-258.
Tan, K.C. and R.J.Bennett, 1984, *Optimal Control of Spatial Systems*, Allen and Unwin, London.

CHAPTER 17

Creative Renaissance of the Osaka Bay Area - Towards a Cosmo-Creative Region in the 21st Century

Kazuhiro Yoshikawa

17.1 WHY BAY AREAS?

Civilization first began and then developed along waterways, rivers and the seas. It was the 'port' that served as the primary gateway to the rest of the world. Through the 'port' passed a variety of commodities such as foods and commercial goods for urban life, as well as artists and craftsmen. It was used by all types of citizens, merchants, politicians, and even soldiers. The 'port' was a meeting and transit point, where a great deal of information was exchanged, and a variety of commodities were traded. It served exactly the same function as areas we now call centres of commerce, culture, and communication.

'Ports' have served a vital historical role. We know that the ancient Greek city-states such as those that developed along the coasts of the Aegean, Adriatic and Mediterranean seas all had a 'port' of their own which was the core of urban activities. In the Medieval Ages in Europe, Venice and Genoa vied with each other to establish the centre of communications between the East and the West. Hamburg, Lübeck, Bremen and other cities successfully formed the Hanse, a medieval league of free port towns. The end of the 15th century saw the great discovery of the American Continent, which triggered what may be called the 'port era' of Europe in the following centuries. It is worth noting that the modern ports of London and Amsterdam are believed to have established a footing by the 17th century. At the time of the Revolutionary War, New York was no more than a small port. The then mayor was very much confident that in order for the city to achieve further growth there would be no more effective measures than the development of its port functions. Thus, he promoted the project of constructing a large canal connecting Hudson Bay and Lake Erie. The canal that was completed in 1825 literally paved the way for the prime importance that the city enjoys today.

All this explain very well how essential it was to develop 'ports', if cities were to grow, and conversely, how critical it was to foster urban activities, if 'ports' were to flourish.

17.2 HISTORICAL PROFILE OF THE OSAKA BAY AREA

In 1859 Japan opened herself again to the rest of the world after the 220 years of isolation policy exercised by the Tokugawa Shogunate. There followed an era of internationalization during which ports became increasingly important. Directly after the Meiji Restoration the government devoted considerable energy to the introduction of modern Western scientific technology in the spheres of manufacturing, mining, transportation, and communications. The degree to which the post-Restoration modern development of the economy could take place depended very much upon the extent to which the West's modern facilities such as steamships, railroads, electric telegraphy, etc., could be introduced. Yet, when judged by Tokugawa period standards, all of these demanded unimaginably vast accumulations of capital. Therefore, initially these facilities could only be introduced through government investment.

It was only around 1887 that merchant shipping and railroads had developed sufficiently as modern transportation facilities and began to make a genuine contribution to the formation of the modern Japanese state. The growth of merchant shipping made a major contribution to the great progress in the development of the modern Japanese economy through the opening of overseas routes, especially by breaking the ground in export trade routes. Before the development of the Japanese merchant marine service, British and American ships controlled most of the foreign routes and quite naturally their shipping lane network and freight charge systems were devised to profit those countries.

When foreign routes were opened up by Japanese ships trade developed rapidly, and the various disadvantages, inconveniences, and irrationalities of the former system were overcome and rectified. The Meiji government quickly recognized the significance of the development of modern port facilities in order to promote overseas trade. Through the introduction of the Western technology of civil engineering, the significant ports of the Tokugawa Era were renovated and transformed into modern ports equipped with trade-supporting facilities. The first such modern port was Yokohama, followed by Ube, Osaka, and so on.

During the postwar period of high economic growth, the Osaka Bay Area has been handling and distributing a growing number of commodities, and has been able to offer extensive space for industrial activities, especially those of heavy industry. The area has thus been contributing to the continued growth of Japan's economy. Extensive areas of reclaimed land, industrial ports dug into the coastline and distribution port terminals for handling international cargoes are all symbols of Japan's ongoing economic growth. The major roles the Osaka Bay Area has performed may be categorized into five distinct functions: (1) a port-based distribution centre, (2) an international gateway for people, culture and information, (3) an industrial location, (4) a centre for urban functions, (5) a leisure/amenity space. The historical development of the area may well be sketched by use of this categorical system (see Table 17.1). Figure 17.1 illustrates the current spatial allocation of the above functions. A mere glance at this figure shows that the development along the waterfronts of the Bay Area is monopolized by industrial and distribution uses. The result is an excessive isolation of the area from citizens' daily urban life. A 'port', in the original sense of the word, connotes a place with a somewhat romantic atmosphere. However, the 'ports' that result from modern engineering may seem rather 'inaccessible to the public' or 'unfriendly to the citizen'. In other words, many of the real waterfronts have been lost to the people and have become totally uninteresting for the general citizen.

Properly speaking, the original meaning of a bay area is a place where both the sea and land meet, or a kind of interface between them. It is a valuable area of land where activities of distribution and industry are conducted, people are working, resting, and living. Undoubtedly, society is now changing rapidly and so are social demands upon port areas. Shifts are occurring from single-function-oriented to multiple-function-oriented areas, from a quantity-orientation to a quality-orientation. In order to allow the Osaka Bay Area to face the beginning of the 21st century in a dynamic way, its future should be designed precisely in line with these tidal changes in society.

17.3 PRODUCT CYCLES IN THE OSAKA BAY AREA: THE OBSOLESCENCE OF HEAVY INDUSTRIES

After its long period of isolation, Japan was plunged into a drastic - for those days - tidal wave of internationalization at the turn of the Meiji era. The Osaka Bay Area was not excluded from this process.

Table 17.1 : Historical Perspective of the Osaka Bay Area

period / roles of the area	Ancient Times (~ 12th century)	Early Feudal Times (12th century ~ 15th century)	Late Feudal Times (15th century ~ 19th century)	Early Modern Times (Meiji. Taisho. Early Showa) (19th century ~ 20th century)	Modern Times (Late Showa. Heisei) (20th century ~)
distribution center	-developed as a trade port -navigation improved -served as an outport to the capital of Japan (Kyoto)	-(trade with the Sun dynasty)	-(trade with the Ming dynasty) → (isolation policy) → (opening the port to foreign countries) - (Osaka-centered distribution system prevailed) - (navigation system improved)	-Kobe Port opened (international) -Osaka Port opened (domestic) -(industrial ports developed) -(navigation system declined)	-international port containerized -(large-scale industrial ports developed) -(navigation system vanished)
gateway to overseas in internationalization. telecommunication. exchanges	-served as a gateway to the Asian Continent -(government missions to the Sui and the Tang dynasties) -(Buddhism introduced)	-served as a gateway to East Asia -(Osaka(Sakai) merchants traded with East Asia) -(Kyoto(Muromachi.Azuchi)-based cultural effervescence)-(Osaka Townsman's Spirit-based cultural effervescence)		-developed as a concession area -served as a cultural gateway to the West -fostered the fashion-conscious port (Kobe Port)	-developing the region as a key station for internationalization -building manmade islands such as the Port Island and the Nanko (the South Port Osaka) -starting the construction of the new international airport
space for industrialization	-fishing industry	-fishing industry	-fishing industry	-fishing industry -developed as an industrial port -coastal industries developed (spinning industry, shipbuilding industry,banks, stock exchange market, trading companies)	-fishing industry -shifting from heavy manufacturing industries to urban industries
space for urbanization	-a community developed	-construction of a big Buddhist temple (Ishiyama Temple) -Central Osaka declined -Southern Osaka(Sakai) developed	-construction of the Osaka Castle -(Osaka revitalized) -(a canal excavated) -agricultural communities developed -a wharf built	-concession areas developed -coastal industrial zones developed	-creating reclaimed lands -(downtown Osaka over-populated) -developing as a high-quality urban area -urban infrastructures improved -port equipped with a contrainerized wharf
recreation/leisure space	-an amusement quarter developed -served as a beauty spot (included in the pilgrim course to Kumano Temple)	-served as a beauty spot	-served as a beauty spot	-an extensive amusement district developed	-serving as a scenic resort -developing into a marine recreation/leisure center
regional backgrounds	-the capital of Japan moved from Osaka through Nara to Kyoto	-the ritual capital of Japan being Kyoto -the capital of Shogunate Government being Kamakura (near Tokyo)	-the ritual capital of Japan being Kyoto -the capital of Shogunate Government being Edo (Tokyo)	-the capital of Japan moved to Tokyo	-Tokyo being the capital of Japanese Government -excessive centralization to Tokyo -Osaka losing power

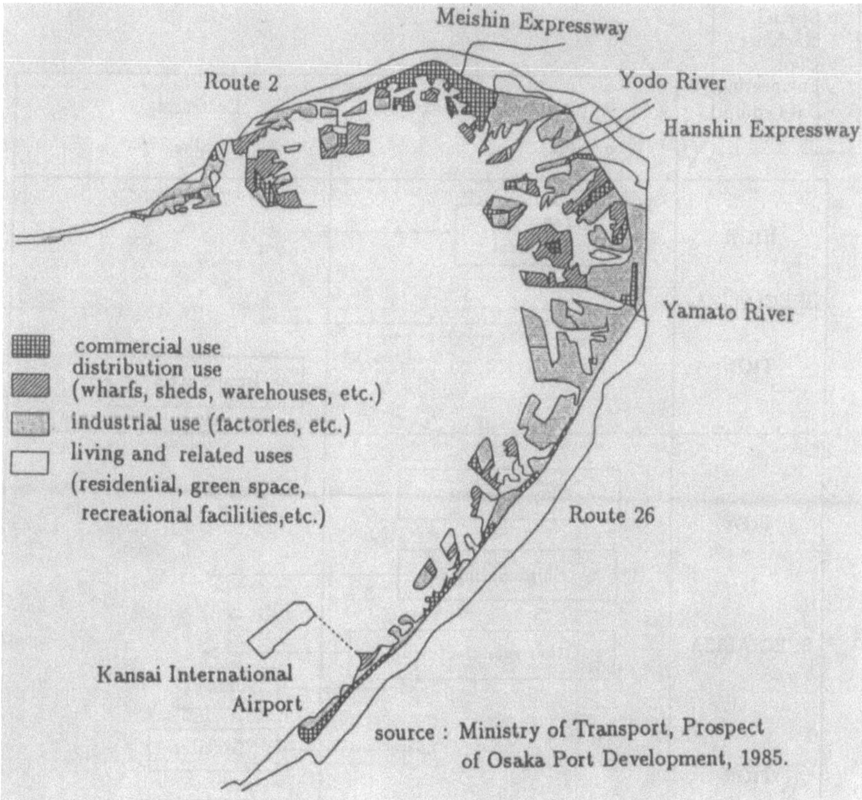

Figure 17.1 : Present land use

It subsequently evolved to meet the needs of the times, introducing the then advanced technology from the West. Since then, Osaka Bay has served as a significant junction between the international and domestic movements of commodities. The roles and functions of the Bay Area were remarkably expanded in response to growing trade and exchanges in the 1960's, a period of rapid economic growth throughout Japan. The area along Osaka Bay provided a vast space for industrial production, especially chemical and heavy industries, which were playing a central role as driving forces of economic growth.

Today, Japan shares the responsibility with other developed countries to pave the way towards a new creative society in the 21st century. The Osaka Bay Area is also expected to become a cosmo-creative metropolis and thus to take on pioneering role in the new era. Thus,

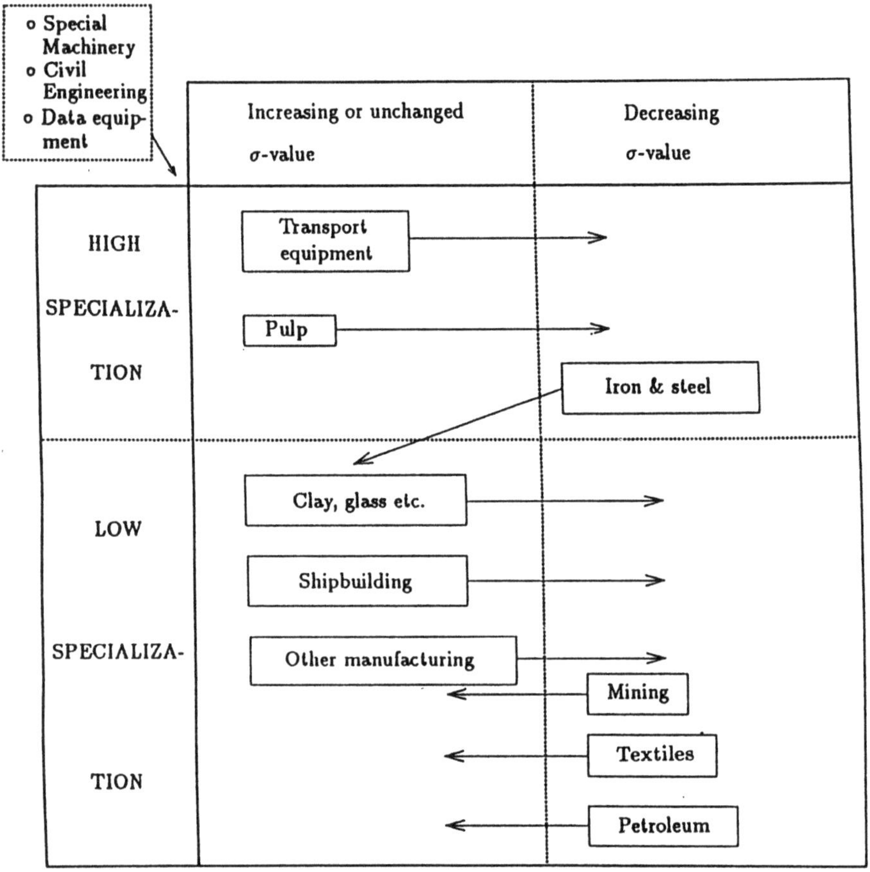

Original source: OECD Trade Statistics 1977-1980.

Cited from: Å. E. Andersson and B. Johansson, Knowledge Intensity and Product Cycles in Metropolitan Regions, IIASA WP-84-13, 1984.

Figure 17.2 : Product categories with changing specialization

the following question may arise: from heavy industry in the 1960's to where? A means for solving this question is to apply the product cycle theory. Figure 17.2 shows the product cycles of the OECD nations (Andersson, 1985).

The second quadrant in Figure 17.2 covers the phase of the advent of new era products. That is, this quadrant represents the specialized products of the OECD nations. R & D activity is a significant indicator of both the levels and availability of knowledge and technology. All of the OECD nations enjoy highly developed infrastructure systems such as transportation networks characterized by high degrees of accessibility. As a particular product technology matures, the share of R & D costs to the total costs decreases and so does the ratio of the number of knowledge-handling workers to the total number of workers. Thus, a highly-matured phase of technology is characterized by a high-degree of capitalization. The steel and metal producing industries are typical examples of this kind of industry. Such industries appear in the first quadrant in Figure 17.2.

At a later stage, when the product is approaching its final form and function, the production units tend to be absorbed, causing great contraction. Industrial policy has to react to international competition. Shipbuilding is a prominent example of an industry which has experienced a very drastic decrease in relative importance in the OECD countries. Most natural resource oriented industries and production units with labor intensive technology relocate almost entirely outside the OECD countries. It takes politicians quite a long time to give up protective policies and heavy subsidies of their domestic industry after perceiving that an industry of this type has been completely taken over by non-OECD nations. The increasing importance of creative investments in different countries of the OECD should be expected to lead to an increased rate of structural change within the next decades, according to the product cycle.

Application of this theory to the Osaka Bay Area indicates that the area should make a shift to knowledge-intensive industries with a high-degree of R & D intensity. In order to accomplish this purpose, a knowledge-oriented environment must be fostered through increased encouragement of face-to-face communications; this in turn requires increased competence and creativity coupled with a focus upon the cultural features of the area. 'Communication forums' for such knowledge activities should be set up. A long-range vision for the Osaka Bay Area during the 21st century should be developed by incorporating the above findings. From the viewpoint of infrastructure development, more integrated spaces for both knowledge-producing and knowledge consumptive activities need to be fostered through comprehensive regional planning.

It is now the high time for the cities in the Osaka Bay Area to pioneer this new logistical revolution, thus leading to increased regional development. This in turn requires the creation of appropriate settings for scientific research and creative activities as well as high-amenity living. That is, creative global cities or what we may call 'cosmo-creative cities' have to be developed in this region. Thus, the Osaka Bay Area is now confronted by what may be called 'a creative renaissance era'. Can it simply be a coincidence that just at this time a variety of projects for redeveloping and revitalizing bay areas is occurring all over the world? London's Dockland, Baltimore's Inner Harbor, New York's South Street Seaport - just to name a few. The Osaka Bay Area cannot afford to isolate itself from this process.

17.4 TOWARDS A COSMO-CREATIVE METROPOLIS

Japan has achieved a high rate of economic growth and now boasts the world's second largest GNP, encompassing 10 % of the world's total GNP. The economy of the Osaka Bay Area enjoys a scale which almost matches that of the total national economy of such developed countries as the Netherlands, Canada and Austria. It is now endowed with considerable international power and exercises great influence on the global economy, while, at the same time, it is also vulnerable to fluctuations in the global economy that affect its very existence. This area has played a central role in the development of various aspects of Japanese culture which have developed through a continuous amalgamation of cultural strands from the Orient and the Occident. The area already possesses some of the necessary structural diversity in its industry, culture, arts and science to become a versatile international node in the cosmo-creative networks of the future.

The qualitative expansion of the regional knowledge base and of accessibility to the most advanced knowledge in the world is a requisite for the creative renaissance of the Osaka Bay Area. This development goes hand in hand with a successive improvement of the air transportation system. In February, 1987, the construction of Kansai International Airport as well as the *Rinku* ('Airfront') Town was started. Since then, much progress has been made; the embankment works have almost enclosed what is to become the Airport Island. The design competition for the terminal building, the most symbolic facility of the airport, has been won by Messrs. Piano and Andrew. The international airport - which is to operate round-the-clock when completed - coupled with the international ports of Kobe and Osaka, will undoubtedly help the area to form a cosmo-creative metropolis. At the same time, the area has to consolidate its functions as a centre for culture, high-quality administration and international exchange and communications.

In line with such a framework, diverse projects are being undertaken (see Figure 17.3) in order to create cores of development by

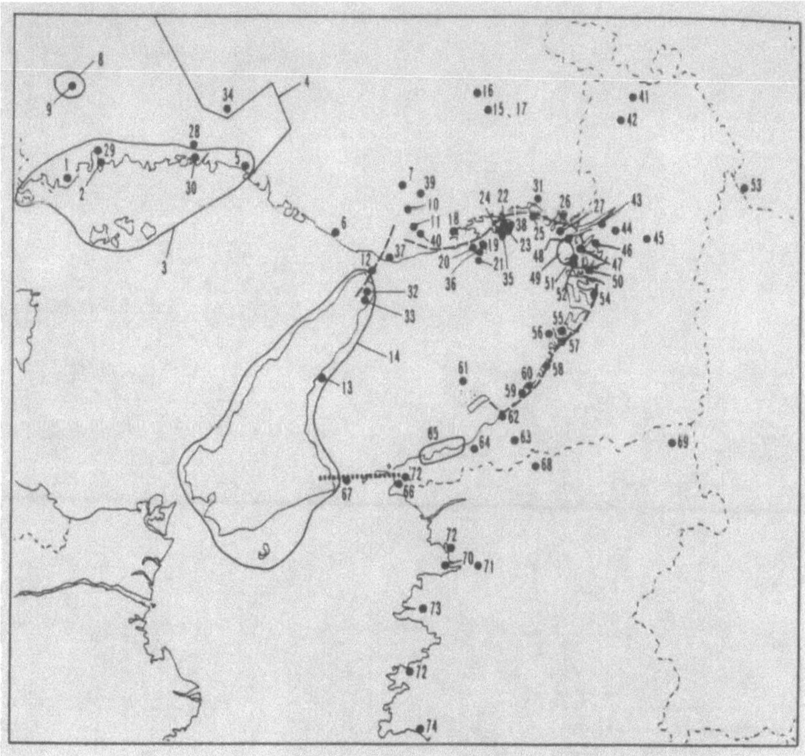

Figure 17.3 : Proposed regional development projects

targeting the beginning of the 21st century. Projects such as the Kansai Cultural and Scientific Research City as well as the Technopolis, Research park, etc., have already been initiated to equip the Osaka Bay Area with knowledge and environment infrastructure, providing a valuable foundation within which dominant factors for the emergence of a cosmo-creative metropolis may be identified. The basic tenet which aligns the various projects being undertaken is the formation of a 'Research Complex', which is expected to catalyze synergetic interactions among creative minds, and, at the same time, to serve as a gateway node open to the global cosmo-creative networks. It is being designed to take full advantage of the communications approaches available in the area; that is, air, sea, ground, and telecommunications routes. In this extremely favorable location as a key source and sink of international traffic flows, regional development projects are now in progress to develop infrastructure for international communication, high-tech industry and leading R & D activities, residential districts

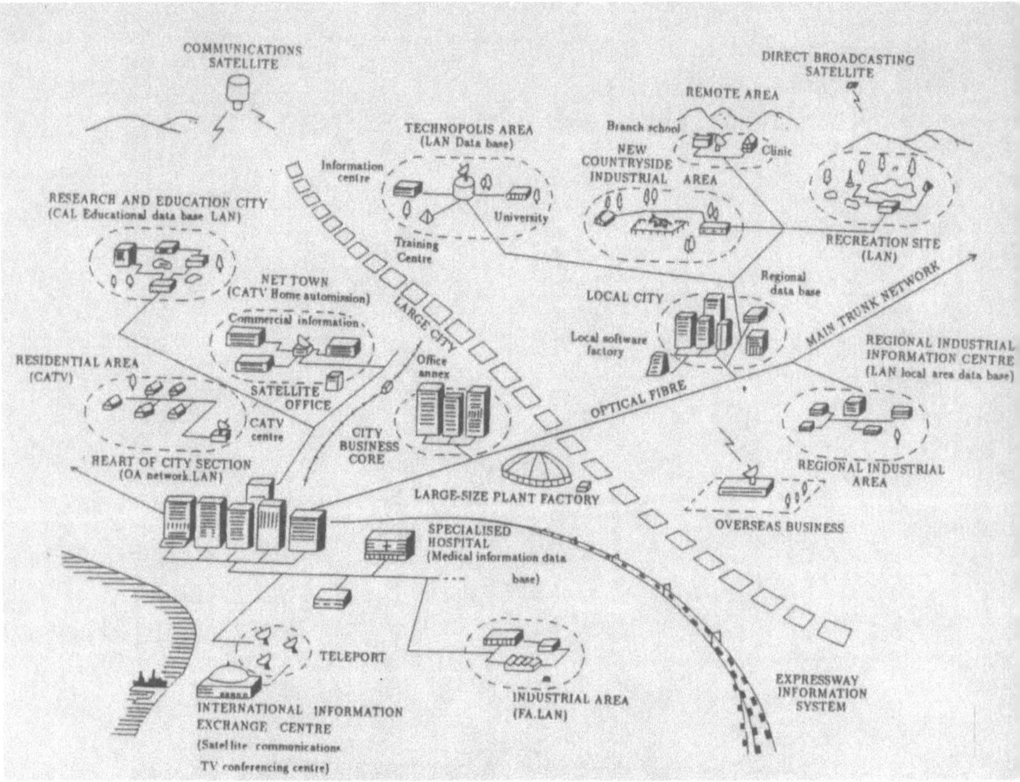

Figure 17.4 : The informized city in the twenty-first century

as well as waterfront-based recreational areas for the citizens.

More specifically, new creative cities are being fostered in the Osaka Bay Area by taking full advantage of the functions of international airports, ports and the 'Teleport'. The Teleport is a ground base for handling expanding international and domestic satellite communications through telecommunications facilities housed in an intelligent building that is linked to other cities by optical fibers and other access circuits (see Figure 17.4). The Teleport is expected to attract heavy users of telecommunications such as financial institutions, software producing companies, information processing services, etc.

A cosmo-creative metropolis can be defined in terms of its activities and functional characteristics. The activities involved may be economic, governmental and socio-cultural. Thus, a city may perform several international functions: as a catalyst of international events such as exhibitions, conventions, etc.; as a research and educational centre focussed upon advanced science and technology; as a focal point of international trade and transactions; as an arbitrator of exchange in the global markets of money, security and commodities; as a key node of transportation and communications networks; and, as an international resort base for marine sports.

17.5 DESIGN OF A GRAND VISION FOR THE FUTURE OF THE OSAKA BAY AREA

As mentioned above, there are many large-scale and long-range projects going on to vitalize the Osaka Bay Area. The most prominent of these developments is the 'Osaka Bay Area Development (OBD)' which refers collectively to a set of challenges for urban and industrial restructuring, especially in the old industrialized zones along Osaka Bay (The Osaka Science and Technology Center, 1991). The goal is to develop a conceptual framework which can integrate the diversified tenets of various projects undertaken in the Osaka Bay Area. The Association for Promoting the Osaka Bay Area Development was established at the initiative of the local business circle, to discuss the grand design for the Osaka Bay Area Development. Their discussions were finally integrated into the Grand Vision, which was announced to the public in April, 1991.

The ideas implicated by the 'cosmo-creative metropolis' are built into the Vision. That is, with a view to nurturing a creative world city complex, a wide range of urban functions is to be consolidated. The Grand Vision envisages the following specific preconditions to bring the cosmo-creative Osaka Bay Area into being (The Association for Promoting the Osaka Bay Area Development, 1991):

(1) the creation of a creatively diversified atmosphere for citizens by amalgamating various urban functions for living, working, learning, and playing;

(2) the formation of cultural and knowledge corridors which bring about synergetic interactions among creative minds;

(3) the formation of an international around-the-clock city for global citizens;

(4) the incubation of new business opportunities and a restructuring of the industrial structure;

(5) the provision of nature-rich amenities for the citizens;

(6) the sophistication of a multi-polar metropolitan structure.

The Grand Vision also envisages that future residents of the Osaka Bay Area be guaranteed at least the following specific life-style attributes.

(1) Attribute 1: 60 minutes-maximum commuting time.

(2) Attribute 2: diversity of choices for living and working environments, including the choice of being in 'walking distance from home to office'.

(3) Attribute 3: let residents be allowed to rely upon multiple places as their 'base cities' or 'mother cities'. For example, on weekdays life a family may live or work in a large big city, but over the weekend they can frequently visit and stay in another city nearby.

(4) Attribute 4: let residents enjoy a variety of cultural and environmental amenities. Waterfronts in the Osaka Bay Area should be made more open to public; environmental quality as well as urban amenities should be improved as a part of infrastructure development.

(5) Attribute 5: we place great importance upon the logistical system which the existing manufacturing industry has built up over a long period so we should encourage the manufacture to restructure their production system by letting them take part in the full-scale spatial restructuring of the Osaka Bay Area. Thus, we should also allow new indsutries to come to the OBA. Secondary and tertiary industries should be encouraged to constitute new urban waterfront zones, part of which should be made available to either residents or visitors.

(6) Attribute 6: The Kansai International Airport should be developed and operated at its maximum potential in order to serve as an around-the-clock, hub international airport. We expect it to be closely interlinked with domestic networks of transportation and to provide a gateway to the international arena.

The Grand Vision states the basic strategies for realizing the cosmo-creative metropolis. The logistical systems required to foster the cosmo-creative metropolis are symbolically indicated by the concept of the 'Osaka Ring Conurbation' (The National Land Agency, 1985).

Other starategies, such as the construction of the Akashi Strait Bridge (the main span is 1990 meters in length) are central to the realization of the ring conurbation by combining the economic and social diversity of Osaka, Kobe and Kyoto with the environmental and recreational richness of Awaji Island and the unique characteristics of Shikoku Island. A proposal is now being studied to connect the channel between Wakayama and southern Awaji. If this eventurates, a loop would be created around the Bay (see Figure 17.5). The public sector may be responsible for providing the finance for such a large-scale hard

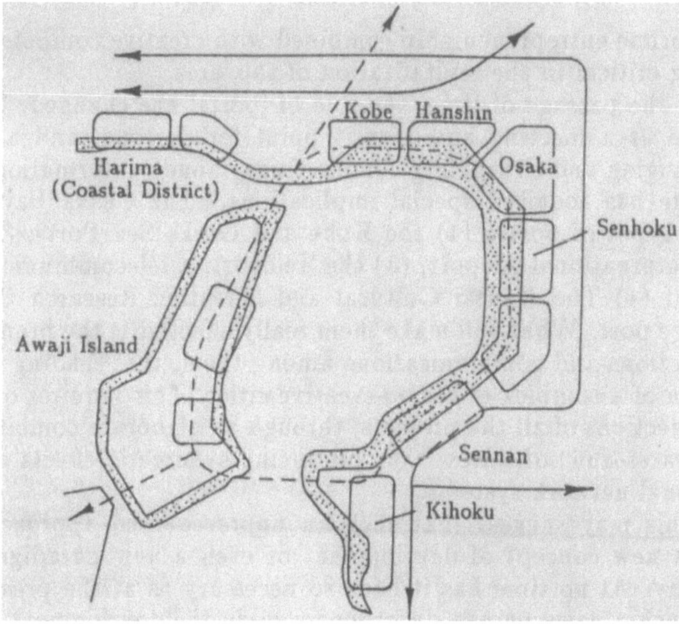

Figure 17.5 : "Osaka Ring Conurbation" Prospect

infrastructure, but each project must also be equipped with an essence of the know-how knowledge accumulated in the private sector.

Thus, the Grand Vision tries to encapsulate the future perspective of the Osaka Bay Area in the cosmo-creative renaissance arena. Instead of formulating precise targets and prescriptions, the Grand Vision serves as the conceptual platform which underpins vastly diversified innovative efforts by public and private sectors intended to create a new type of cosmo-creative metropolis. This conceptual platform is the indispensable infrastructure for innovative planning exercises.

17.6 CONCLUSION

As it has been repeatedly pointed out, the expected role of the Osaka Bay Area in the 21st century is to serve as a complex of creative internationally-oriented cities based on the characteristics of a cosmo-creative metropolis. To accomplish this goal, centres for knowledge production are urgently required. No less vital is the further encouragement of prime entrepreneurship - the powers, spirits and intelligence to transform knowledge into production. As Walt Disney put it, 'if we can dream it, we can do it' - this is how he made the business of leisure the gigantic industry it is today. For the first time in history, prime entrepreneurship combined with creative competency is becoming critical to the revitalization of this area.

With the passage of time, the role of 'ports' has changed. Today their role as a meeting and transit point for persons, and a place for exchanging and channeling commodities, money, information, and knowledge has acquired special implications. The Osaka Bay Area has four types of ports: (1) the Kobe and Osaka Sea Ports, (2) the Kansai International Airport, (3) the Teleport, a telecommunications port, and (4) The Kansai Cultural and Scientific Research City, a knowledge port. What will make them really effective is the promotion of interactions and communications among them, thus leading to the formation of a complex of cosmo-creative cities. This requires organic interconnections of all the projects, through an elaborate combination of hardwares and softwares, thus producing synergetic effects among the regional network systems.

All this may suggest that such an unprecedented approach will require a new concept of development or even a new paradigm and philosophy. At no time has it been so necessary as at the present to pursue such a new concept, in order to guide the development of the Osaka Bay Area during the 21st century in the best way!

REFERENCES

Andersson, Å.E., 1985, *Creativity: The Future of the Metropolis*, Prisma, Stockholm (in Swedish).

The Association for Promoting Osaka Bay Area Development, 1991, *Grand Design of Osaka Bay Area Development*, The Association for Promoting Osaka Bay Area Development, Osaka (in Japanese).

The National Land Agency, 1985, *Kokudocho Keikaku*, The National Land Agency, Tokyo (in Japanese).

The Osaka Science and Technology Center, 1991, "The Osaka Bay Area and Its Industrial Development-Plan 21", Working Report on the Future Prospect of Industries in the Osaka Bay Area and Possible Future Evolutions, The Osaka Science and Technology Center, Osaka.

LIST OF CONTRIBUTORS

Åke E. Andersson
Institute for Futures Studies
Hagagatan 23B, 3tr.
113 47 Stockholm
Swedcn

Christer Anderstig
Office of Regional Planning and Urban Transportation
Stockholm County Council
Box 12557
102 29 Stockholm
Sweden

David F. Batten
Department of Regional Planning
Royal Institute of Technology
100 44 Stockholm
Sweden

Hermann Haken
Institute for Theoretical Physics
University of Stuttgart
Phaffenwaldring 57/111
7000 Stuttgart
Germany

Eizo Hideshima
School of Civil Engineering
Kyoto University
Kyoto 606
Japan

Björn Hårsman
Office of Regional Planning and Urban Transportation
Stockholm Conty Council Office
Box 12557
102 29 Stockholm
Sweden

Börje Johansson
Department of Regional Planning
Royal Institute of Technology
100 44 Stockholm
Sweden

Masuo Kashiwadani
Department of Civil and Ocean Engineering
Ehime University
Matsuyama 790
Japan

Kiyoshi Kobayashi
Department of Social Systems Engineering
Tottori University
Tottori 680
Japan

Se-il Mun
Graduate School of Information Sciences
Tohoku University
Ojoken Katahira 2
Sendai 980
Japan

Norio Okada
Disaster Prevention Research Institute
Kyoto University
Uji 611
Japan

Makoto Okumura
School of Civil Engineering
Kyoto University
Kyoto 606
Japan

Donald G. Saari
Department of Mathematics and of Economics
Northwestern University

Evaston, Illinois 60208-2730
USA

Komei Sasaki
Graduate School of Information Sciences
Tohoku University
Ojoken Katahira 2
Sendai 980
Japan

Gudmund J. W. Smith
Department of Psychology
Lund University
22350 Lund
Sweden

Seishin Sunao
Chuo-Fukken Consultants Co. Ltd.
Higashi-Mikuni
Osaka 532
Japan

Kazuhiro Yoshikawa
School of Civil Engineering
Kyoto University
Kyoto 606
Japan

Wei-Bin Zhang
Institute for Futures Studies
Hagagatan 23B 3tr.
113 47 Stockholm
Sweden

MIX
Papier aus verantwortungsvollen Quellen
Paper from responsible sources
FSC® C105338

FSC
www.fsc.org

If you have any concerns about our products,
you can contact us on
ProductSafety@springernature.com

In case Publisher is established outside the EU,
the EU authorized representative is:
Springer Nature Customer Service Center GmbH
Europaplatz 3, 69115 Heidelberg, Germany

Printed by Libri Plureos GmbH
in Hamburg, Germany